SEA & LAKE MONSTERS Wonders & DEEP SECRETS

A MIND-BLOWN™ SERIES FUN BOOK

1000+ Fun & Weird Facts, Myths, Legends, Stories +

THE ULTIMATE FUN, WEIRD AND WONDERFUL BOOK ON THINGS THAT ARE DEEP, UNKNOWN, SCARY and WOW!"

MIND-BLOWN

ISBN: 979-8-9938493-5-5

MIND-BLOWN™ PUBLISHING TEAM

IN COLLABORATION WITH WG COAKLEY PUBLISHING
ILLUSTRATIONS: LUMEN GARY, ALDEN SKETCH
RESEARCHERS: HOLLIS SCOUT, WILDER FACTUM

ICON ATTRIBUTION TO: FREEPIK – FLATICON.COM, ADOBE.COM, GOOGLE
GRAPHIC ASSISTANCE BY: OPENART.AI, MIND-BLOWN CARTOONS

PART OF THE MIND-BLOWN™ BOOK SER
FIRST EDITION
PRINTED IN THE UNITED STATES

The Official Stamp of
Awesome Weirdness

This is <u>NOT</u> another sleepy textbook.

This is a fun MIND-BLOWN^{tm} book.

Inside, you'll find:

Fun & Weird Facts - quick-hit mind-benders

Myths — Busted - backyard myths crushed by real science

Legends - eerie, maybe-true tales from the deep

Did You Know? - deeper insights into the abyss

Story Moments - short scenes — no towel needed

Fun Quizzes - fast challenges to test your knowledge

Jokes & Comics - Mind-Blown™ weird and slippery

Games with QR Codes

Strange Creatures, Wonders, and Secrets of the Deep.

Real science style images.

CONTENTS

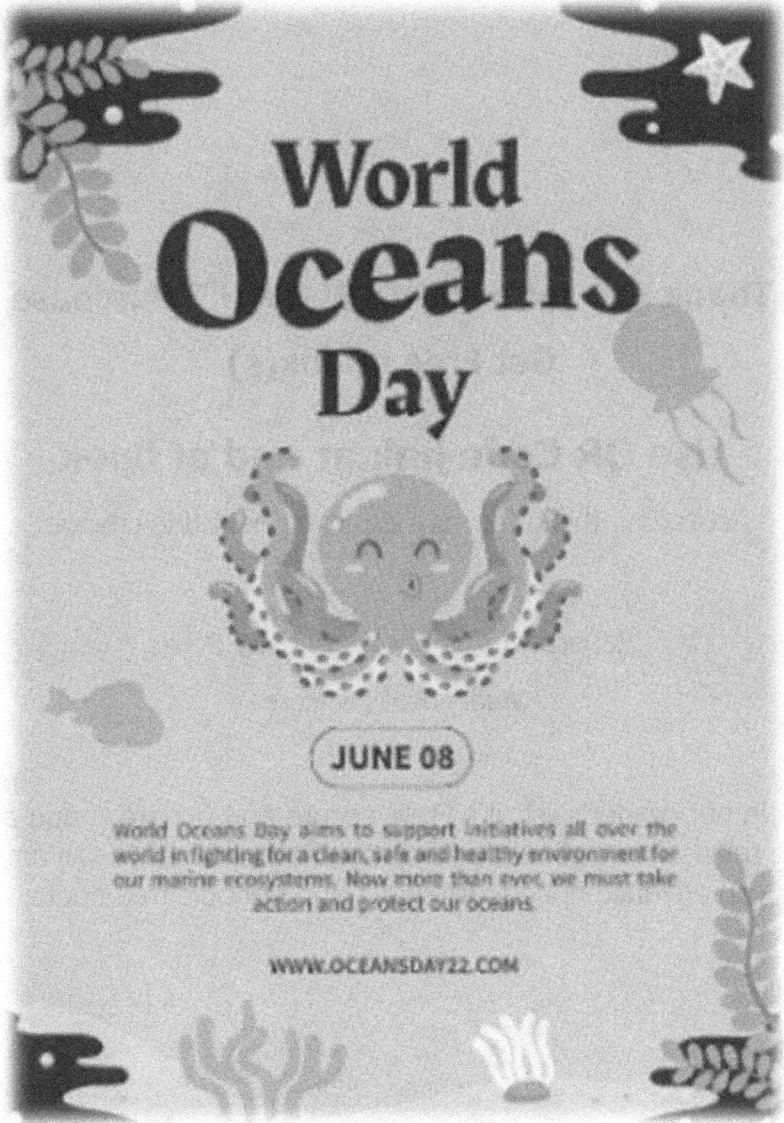

CHAPTER 1

1-GIANTS BELOW
When Size Stops Making Sense

Giant sea squids were known for centuries only through scars on whales and washed-up remains. Living specimens were unseen until modern cameras reached great depths.

FUN & WEIRD FACTS

MIND-BLOWN MOMENT

MB Fact = Mind-Blown Fact
Real discoveries. Real mysteries, Real Facts. Fun to make you think.

MB FACT: The **blue whale** is so massive that its heart alone weighs about as much as a small car. Its heartbeat can be detected from miles away underwater. Size at this scale changes how biology works.

MB FACT: Some underwater environments produce animals that grow far larger than their shallow-water relatives, even though food is scarce. In stable, cold conditions, bodies grow slowly but continuously without seasonal interruption. Predators are fewer, competition is reduced, and growth has no natural stopping point. Over decades or centuries, size accumulates almost unnoticed. In the deep, patience builds giants.

MB FACT: Giant ocean isopods, distant relatives of pill bugs, can reach the length of a human forearm. They live on deep seafloors across the Atlantic and Pacific, where food may arrive only a few times per year. When a large carcass sinks, they gorge and store energy internally. Their size allows them to survive long famine periods without constant feeding. Bulk becomes a survival reserve.

MB FACT: Extreme pressure favors flexibility rather than rigid strength. Many large deep-water animals have soft, gelatinous, or cartilage-based bodies instead of hard skeletons. Pressure compresses evenly, allowing flexible tissues to survive where stiff structures crack. This is why some massive animals feel surprisingly delicate. Underwater strength often looks like softness.

MB FACT: gray whale migrate farther than any other mammal on Earth. Some travel over 12,000 miles round trip every year. Navigation at that scale suggests sensory abilities humans still don't fully understand.

MB FACT: Giant tube worms near hydrothermal vents grow without mouths or digestive systems. They rely on internal bacteria that convert toxic chemicals into usable energy. This constant chemical supply allows uninterrupted growth. Over time, steady energy production leads to extreme size. These animals prove eating isn't required for growth.

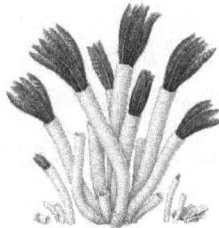

MB FACT: More than 80% of Earth's ocean floor has never been mapped in high detail. Vast mountains, trenches, and valleys exist beyond the reach of modern charts.

MB FACT: The deepest known point in the ocean is deeper than Mount Everest is tall. If Everest were dropped into the Challenger Deep, its peak would still be underwater.

MB FACT: The deep ocean is the largest habitat on Earth. It covers more area than all forests, deserts, and grasslands combined.

MB FACT: sperm whale dive over a mile deep into total darkness to hunt giant squid. At those depths, pressure would crush most machines. Their heads are filled with oil that helps control buoyancy like a living submarine.

MB FACT: The **colossal squid** has the largest eyes ever measured in the animal kingdom. These eyes are not built for sharp detail, but for detecting faint movement in near-total darkness. A passing shadow can signal danger long before contact occurs. In the deep ocean, seeing motion matters more than seeing color. Size shifts from appearance to detection.

MB FACT: Some deep-sea crabs grow thicker bodies instead of longer legs. Increased mass helps them resist pressure changes and survive infrequent meals. Shape evolves around stability rather than speed. In harsh environments, durability matters more than agility. Evolution trades elegance for reliability.

MB FACT: Humans are extremely poor judges of size underwater. Darkness removes reference points, while artificial lights cast exaggerated shadows. Even trained scientists routinely misjudge scale at depth. Many early reports of "impossible size" were honest perception errors. The ocean quietly distorts reality.

MB FACT: Japanese spider crabs can span wider than a small car, yet take decades to reach full size. Found in deep coastal waters near Japan, they grow slowly year after year. Their size is not explosive growth but long-term accumulation. Time itself becomes the growth engine. Longevity rewards patience.

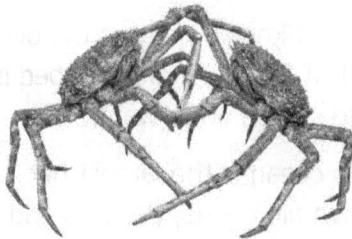

MB FACT: Some underwater animals grow large because they reproduce less frequently. Fewer offspring mean more energy goes into individual growth. This strategy favors survival in stable environments with low predation. Quality replaces quantity. Bigger young have better odds.

MB FACT: Large size underwater often replaces speed as a defense. Fast movement burns energy that may not be replaced for months or years. Many giants remain still for long periods, conserving resources. Stillness reduces risk. In the deep, motion attracts attention.

MB FACT: Glass sponges can grow into massive reef-like structures and live for centuries. Found in cold, deep waters, they grow millimeter by millimeter over time. Their bodies quietly record environmental history. Some living specimens began growing before modern nations existed. Size becomes a timeline.

MB FACT: Submerged mountains and ridges create feeding highways for deep-water life. Currents deliver organic material along predictable paths. Animals living there can grow larger due to reliable supply. Geography shapes biology. Location matters as much as species.

MB FACT: Some deep-water fish grow slowly but live extraordinarily long lives. Low metabolism and stable conditions reduce cellular damage. Extended lifespan allows continued growth. Aging underwater looks different than aging on land. Time stretches biology.

MB FACT: beaked whales are so rarely seen that new species are still being discovered in the 21st century. Some hold the record for the deepest and longest mammal dives ever recorded. Humans know less about them than the Moon's surface.

MB FACT: Giant amphipods grow far larger at depth than their shallow cousins. Pressure affects how oxygen dissolves and is absorbed. This alters metabolic efficiency. Chemistry quietly changes size potential. Physics shapes biology.

MB FACT: Fossil records show deep-water gigantism existed millions of years before humans. Extreme size is not a modern phenomenon. It predates pollution, climate change, and industrial fishing. Giants are ancient residents. Humans arrived late.

MB FACT: Many underwater giants are scavengers rather than predators. They survive on whatever sinks from above. Size allows them to dominate rare food events. Power comes from endurance, not aggression. Waiting wins.

MB FACT: Blue whales, the largest animals ever known, are surface-feeders that rely on dense food swarms rather than constant meals. Their size allows efficient long-distance travel between feeding grounds. Even familiar giants follow the same energy rules. Bigger bodies survive feast-and-famine cycles.

MB FACT: Some underwater animals only reach giant size at depth, remaining smaller in shallow water. Environmental pressure controls growth potential. Biology responds directly to surroundings. Same species, different size rules. Context matters.

MB FACT: Sound travels farther underwater than in air, especially at low frequencies. Large bodies displace more water and generate stronger acoustic signals. This allows giants to sense distant movement. Hearing can replace sight. Sound becomes scale.

MB FACT: Some deep-sea creatures create their own light using bioluminescence. This glow can attract prey, confuse predators, or even make animals invisible from below.

MB FACT: Sound travels farther underwater than in air. Certain low-frequency noises can cross entire ocean basins without fading.

MB FACT: Entire mountain ranges run across the ocean floor. The Mid-Ocean Ridge stretches around the planet like a submerged spine.

MB FACT: Early sonar operators often dismissed massive returns as equipment errors. Objects appeared too large to be believable. Only repeated readings confirmed reality. Technology had to catch up to biology. The ocean kept exceeding expectations.

MB FACT: Giant squids were known for centuries only through scars on whales and washed-up remains. Living specimens were unseen until modern cameras reached depth. Absence of evidence delayed acceptance. Reality existed long before proof.

MB FACT: Some freshwater environments mimic deep-sea conditions closely enough to produce unusual size. **Lake Baikal amphipods** grow far larger than typical freshwater species. Depth, cold, and stability matter more than salt content. Categories blur underwater.

MB FACT: Large animals are easier to detect on sonar but harder to visually identify. Technology exaggerates presence while obscuring detail. This fuels mystery. Size becomes known without form.

MB FACT: Growth underwater often occurs in bursts following rare feeding events. Long dormancy alternates with sudden expansion. Biology adapts to unpredictability. Waiting is part of the design.

MB FACT: Many underwater giants show little reaction to human machines. Lights, sound, and movement often go ignored. Indifference suggests confidence. Curiosity is optional when survival is assured.

MB FACT: In total darkness, silhouette replaces color as a signal. Large shapes dominate perception. Size becomes visibility. Darkness rewrites communication.

MB FACT: Some deep-water animals grow larger simply because nothing forces them to stop. Without seasonal stress, growth continues quietly. Nature experiments when left undisturbed. Giants are the result.

MB FACT: Submersible pilots often describe scale shock after encounters with large animals. Familiar reference points vanish. Perspective collapses. Reality feels unreal.

MB FACT: The deeper humans explore, the more size expectations fail. Each expedition recalibrates "normal." Discovery keeps expanding boundaries.

MB FACT: The **Loch Ness Monster** legend didn't explode because of medieval myths -- it went viral after a single blurry 1934 photo. That image permanently rewired how humans interpret dark water shapes. Once an idea exists, the brain *expects* monsters.

MB FACT: Loch Ness holds more freshwater than *all* the lakes in England and Wales combined. Its depth, peat-dark water, and cold temperatures create near-perfect conditions for hiding large, moving objects. Visibility can drop to just a few feet.

MB FACT: The **Kraken** may have been inspired by real **giant squid or octopus** -- creatures so large they can tangle with whales. For centuries, sailors reported tentacles longer than ships *before* science confirmed the species existed.

MB FACT: giant squid have eyes the size of dinner plates -- the largest in the animal kingdom. That's not for drama; it's for detecting movement in near-total darkness. Deep oceans reward paranoia.

MB FACT: Japan's **Umibōzu** legends describe black shapes rising suddenly from calm seas to smash ships. Modern oceanographers note that rogue waves can form *without warning*, even on clear nights. The ocean doesn't need myths to be lethal.

MB FACT: There are underwater "waterfalls" taller than skyscrapers. These occur when dense cold water spills over submerged ridges and plunges downward.

MB FACT: New deep-sea species are discovered almost every time scientists explore extreme depths. Many are seen only once and never encountered again.

MB FACT: Some deep-sea animals may live hundreds of years. Cold temperatures and slow metabolisms dramatically slow aging.

MB FACT: Sailors worldwide describe "something following the ship" just below the surface. This sensation often comes from bioluminescent plankton reacting to pressure waves. Your brain interprets the glow as pursuit.

MB FACT: Many "sea serpent" sightings match the movement of **oarfish**, a real fish that can exceed 30 feet in length. It lives deep, rises rarely, and looks absolutely unreal at the surface. Legends love rare appearances.

Beaked whales

Beaked whales are a diverse family of deep-diving cetaceans (family **Ziphiidae**) known for their elongated beaks and elusive behavior. They inhabit deep ocean waters worldwide and are among the least understood large mammals because they spend little time at the surface and dive to extreme depths for long periods.

These whales are highly specialized deep divers, foraging in meso- and bathypelagic zones where light barely penetrates. They use echolocation clicks to hunt squid and small fish. Surface behavior is inconspicuous -- they often rise silently and spend only minutes above water -- making them hard to detect in surveys.

FACT: Deep-sea anomalies often trigger intense scientific debate and oversized imaginations.

FACT: Giant squids (*Architeuthis dux*) can grow up to 43 feet, but their culinary skills remain unconfirmed.

FACT: The deep ocean is a realm of darkness, pressure, and occasionally, things that look distressingly like produce.

FACT: Lake bottoms are often repositories for human history, including misjudged aquatic vehicle disposals.

FACT: Bioluminescence is common in deep-sea organisms, sometimes confusing them with much larger threats.

FACT: We study the ocean's mysteries, but sometimes, the ocean's inhabitants are studying us right back.

MIND-BLOWN™ Cartoons

Fun Fact Comics

MIND-BLOWN™ Cartoons

MYTHS - BUSTED

MYTH: Giant underwater creatures are aggressive monsters waiting to attack ships or submarines. This belief grew from dramatic art and fear-based storytelling. In reality, most large underwater animals avoid unnecessary contact. Attacking wastes precious energy. Size underwater is about survival, not violence.

MYTH: Bigger animals must be stronger than smaller ones. Extreme pressure favors flexibility rather than brute force. Many giants have soft or delicate bodies that would be damaged by sudden motion. Strength underwater looks different. Endurance matters more than muscle.

MYTH: Huge underwater animals need to eat constantly. Scarcity is exactly why many of them grow large. Bigger bodies store energy longer and survive extended famine. Some species go years between major meals. Size is insurance against uncertainty.

MYTH: Sightings of massive underwater creatures are just exaggerations or hallucinations. While perception errors exist, many reports came from trained observers. Modern cameras and sonar have confirmed animals once thought impossible. Skepticism delayed discovery more than imagination. Humans underestimated the ocean.

MYTH: Freshwater environments cannot produce large or unusual life. Certain lakes are deep, cold, and stable enough to mimic ocean conditions. Lake Baikal proves this repeatedly. Environment shapes size, not location labels. Nature ignores categories.

LEGENDS

REAL LEGENDS

The Long Shadow off Cape Finisterre
North Atlantic -- Spanish and Portuguese sailors

Sailors rounding Cape Finisterre spoke of a shadow that paced their ships on calm nights. Lantern light revealed movement beneath the hull, darker than the surrounding sea. The shape stretched longer than the vessel itself and matched its speed precisely. Crews fell silent as it followed, then slowly dropped back and vanished. Logs warned captains not to linger after dark, noting that "the water sometimes follows."

The Sound Beneath the Ice
Arctic Ocean -- 19th-century exploration crews

Ships trapped in winter ice reported deep, rhythmic sounds traveling through hull and deck. The noise was not cracking ice, but slow pulses felt more than heard. Men described the unsettling sense of being observed from below. The sound would fade without surface disturbance. Later theories pointed to massive animals moving under ice sheets, but crews remembered the silence that followed most.

The Still Shape of Lake Baikal
Siberia -- local divers and fishermen

Divers descending into Baikal's depths reported encountering large, motionless forms suspended in black water. These shapes did not react or stir sediment. Fishermen avoided areas where nets returned damaged without obvious cause. Scientists suggested optical distortion and depth effects. Locals still say the lake hides size in stillness.

The Net That Rose on Its Own
Sea of Japan -- coastal fishing villages

Fishermen hauling deep nets sometimes felt them lift upward against their boats. The pull was steady, not violent. When surfaced, nets were stretched or torn, yet empty. Villagers marked those waters and avoided them seasonally. Some said the net wasn't snagged -- it was raised.

The Shape That Outpaced the Wake
Southern Ocean -- whaling logs

Whaling ships recorded long forms moving just beneath the surface faster than the ship's wake. The sea remained calm, yet something paced them silently. No breach followed. No attack came. After prolonged escort, the movement vanished. Crews recorded the event without explanation.

Fun Fact Comics

Fun Fact Comics

An old-world style map inspired by early explorers' charts, marking the seas and regions where legendary sea monsters were once believed to dwell. From krakens and sea serpents to ancient creatures of myth, it reflects how mystery and imagination filled the unknown waters.

DID YOU KNOW ?

Did You Know ? Humans evolved to judge size, distance, and motion in air, not water. Underwater, light bends, shadows stretch, and reference points disappear. Even trained observers misjudge scale dramatically. Many historical size reports were honest perception failures. The ocean rewires how the brain interprets reality.

Did You Know ? Deep-water environments are among the most stable on Earth. Temperatures remain nearly constant and sunlight never reaches them. Without seasonal stress, growth continues uninterrupted. Over decades or centuries, this leads to extreme size. Stability breeds giants.

Did You Know ? Early sailors lacked sonar, GPS, and reliable depth measurement. Calm water amplified movement beneath ships. Darkness magnified uncertainty. When something followed silently, imagination filled gaps technology later clarified. Legends often began with real events.

Did You Know ? Large underwater animals often replace speed with endurance. Fast movement wastes energy in food-poor environments. Slow motion conserves resources and reduces detection. Patience becomes a survival trait. Time favors the calm.

Did You Know ? Sound behaves differently underwater, traveling farther and faster than in air. Low-frequency vibrations pass through water, ice, and hulls. This explains mysterious sounds reported by crews. Physics amplifies presence.

Loch Ness "Monster"

Rising from the dark waters of Scotland's Loch Ness, *Nessie* has been reported for more than **1,500 years**, with the earliest account dating to **565 AD**. Loch Ness stretches nearly **23 miles long**, plunges over **750 feet deep**, and holds more freshwater than all the lakes in England and Wales combined -- creating vast, shadowy depths where large shapes can disappear in seconds.

Since the first modern sighting in **1933**, more than **1,000 eyewitness reports** and unexplained sonar detections have kept the mystery alive, making Nessie one of the world's most enduring legends.

Nessie

STORY MOMENT

Something That Didn't Need to Hurry

The research vessel drifted quietly above the black water, engines barely humming. Below, the remotely operated vehicle descended, its lights cutting narrow tunnels through darkness. The seafloor appeared empty, featureless, calm.

Then a shadow crossed the screen.

At first, the pilot blamed the camera angle. But the shape moved steadily, smoothly, without disturbing the sediment below. It crossed the frame without reacting to light or sound. Someone leaned closer. No one spoke.

The form was enormous, longer than the vehicle, longer than the ship above. Yet it showed no interest in the machine observing it. It didn't change speed. It didn't turn.

For nearly a minute, it passed beneath them.

When the screen returned to empty water, the room stayed silent. The footage replayed again and again. No measurement fit cleanly. It wasn't hunting. It wasn't fleeing.

Later, the expedition lead wrote one line in the log:

"Whatever that was, it had already been there long before us , and it had nowhere else it needed to be."

The ocean closed back around the empty image, unchanged.

FUN QUIZ

1. What term describes unusually large animals living at depth?
   A) Abyssal mutation
   B) Deep-sea gigantism
   C) Pressure growth
   D) Mega-evolution

2. Why does extreme depth favor flexible bodies?
   A) Reduced gravity
   B) Faster movement
   C) Pressure compression
   D) Warmer temperatures

3. True or False: Large underwater animals are always aggressive predators.

4. Which factor most encourages extreme size underwater?
   A) Bright sunlight
   B) Stable environments
   C) Fast reproduction
   D) Shallow water

5. Why is size hard to judge underwater?
   A) Light distortion
   B) Lack of reference points
   C) Darkness
   D) All of the above

FUN QUIZ ANSWERS

1. B
2. C
3. False
4. B
5. D

CHAPTER 2

2-SILENT WORLDS
Life Where Light Never Reaches

FUN & WEIRD FACTS

MB Fact = Mind-Blown Fact
Real discoveries. Real mysteries, Real Facts. Fun to make you think.

MB FACT: Vast regions of Earth exist in permanent underwater darkness where sunlight has never reached. These zones are not empty voids but fully functioning ecosystems powered by chemistry instead of light. Energy comes from heat, minerals, and slow reactions rather than photosynthesis. Life adapts to sense rather than see. Darkness becomes a stable environment, not a disadvantage.

MB FACT: In total darkness, color loses its purpose. Producing pigment costs energy, so many deep-water animals evolve pale, red, or translucent bodies. Red light disappears first underwater, making red animals appear black and invisible. What looks eerie to humans is simply efficiency. Darkness rewrites camouflage rules.

MB FACT: Anglerfish survive where light never reaches by producing their own. Their glowing lure is powered by symbiotic bacteria, not the fish itself. The light attracts curious prey in a world where motion is otherwise undetectable. In darkness, light is deception, not illumination. Seeing the glow often means danger is already close.

MB FACT: The *blobfish* and many deep-sea rattail fish don't rely on vision at all. Living thousands of feet down where light never reaches, they sense tiny pressure changes and vibrations instead. In the deep, movement itself becomes a message.

MB FACT: At depths below sunlight, animals don't experience day or night. Species like deep-sea crabs and fish become active only when food arrives -- such as sinking debris or dead animals. With no sunrise or sunset, life runs on opportunity, not time.

MB FACT: The *vampire squid* lives in oxygen-poor darkness nearly 3,000 feet deep, where most animals would suffocate. Instead of chasing prey, it drifts slowly and feeds on falling organic particles. Its dark red body looks black in deep water, hiding it from predators. Despite its name, it rarely attacks anything.

MB FACT: The *giant squid* has some of the largest eyes in the animal kingdom, each as big as a dinner plate. These eyes don't see color or detail -- they detect faint movement in near-total darkness. In places with no light at all, other species lose their eyes entirely. Evolution keeps only what works.

MB FACT: More than 90% of deep-sea animals produce their own light. Creatures like anglerfish, jellyfish, and lanternfish glow to lure prey, signal mates, or confuse enemies. Some shine constantly, while others flash like underwater Morse code.

MB FACT: Octopuses, even familiar species, behave very differently in deep darkness. Without visual cues, they rely on touch and chemical sensing. Their arms constantly explore their surroundings. Intelligence adapts to sensory limits. Thinking does not require sight.

MB FACT: Darkness slows change. Without sunlight driving rapid biological cycles, deep ecosystems remain stable for thousands of years. Growth is slow and evolution is patient. Some habitats change less in centuries than surface environments do in decades. Time stretches where light never arrives.

MB FACT: Humans experience darkness as fear because vision dominates our senses. Underwater life experiences darkness as normal. Anxiety is a surface habit, not a biological rule. The deep is not hostile -- it is unfamiliar. Fear often says more about us than the environment.

MB FACT: Sound travels farther underwater than in air, especially at low frequencies. Animals rely on vibrations to detect distant movement. In darkness, hearing replaces sight over long distances. Silence carries information. Sound becomes space.

MB FACT: Some deep-sea animals glow only when disturbed. Sudden flashes startle predators and draw attention away from escape. Light becomes a defensive weapon. Surprise matters more than strength. Darkness favors tactics over force.

MB FACT: Lanternfish live mostly in darkness and rise only at night. Their daily migration moves more biomass than any other animal movement on Earth. This happens largely unseen. Darkness hides one of the planet's largest biological processes. Size isn't always visible.

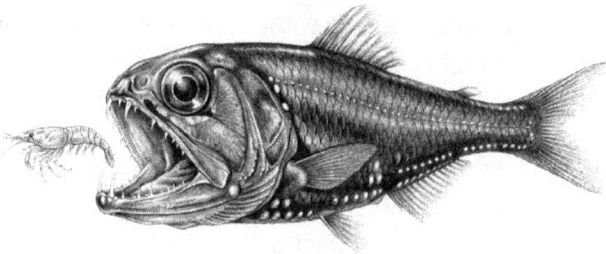

MB FACT: The *blind cave fish* of Mexico's underwater caves are born, live, and die without ever seeing light. These fish hatch with eyes, but the eyes stop growing and disappear because darkness makes them useless. They navigate by sensing tiny water movements instead of sight.

MB FACT: The *deep-sea anglerfish* spends most of its life between 3,000 and 10,000 feet deep, where sunlight never reaches. It produces its own glowing lure to attract prey in complete blackness. The light isn't for seeing -- it's for hunting.

MB FACT: In underwater caves near Florida and the Caribbean, *cave shrimp* and *cave crayfish* live in permanent darkness just a short distance from land. Like deep-sea animals, they are pale, slow-moving, and often eyeless. Darkness -- not distance from shore -- changed them.

MB FACT: The *giant squid* lives so deep and far from light that humans did not film one alive until 2004. Its eyes are the size of dinner plates, built to detect faint movement, not color. It sees shadows, not shapes.

MB FACT: At hydrothermal vents nearly 8,000 feet down, *tube worms* live without sunlight, mouths, or stomachs. Instead of eating, they rely on bacteria inside their bodies that turn chemicals from the Earth into food. Life runs on chemistry, not light.

MB FACT: Some deep-sea fish, like the *viperfish*, flash light from their bodies to communicate in total darkness. These signals can warn enemies, attract mates, or confuse prey. In the deep ocean, light is a language.

MB FACT: When a whale dies and sinks to the deep ocean floor, it creates a "whale fall" that feeds creatures for decades. In total darkness, worms, crabs, and bacteria swarm the bones, turning death into a long-lasting ecosystem.

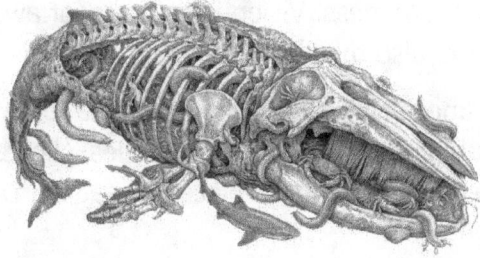

MB FACT: In the deepest parts of the ocean, animals do not follow day or night cycles. Instead, they wait for rare events -- falling food, shifting currents, or chemical signals -- to trigger movement and reproduction. In the dark, timing replaces daylight.

MB FACT: Some deep-sea animals glow blue because blue light travels farthest underwater. Color choice follows physics. Biology adapts to wavelength. Light becomes calculation. Nature optimizes.

MB FACT: In darkness, boundaries disappear. Animals drift without visual landmarks. Navigation relies on memory and current flow. Space feels endless. Direction becomes relative.

MB FACT: Submersibles often detect life in darkness before seeing it. Sonar, vibration, and chemical sensors register presence first. Vision confirms last. Humans lead with the wrong sense. Darkness flips priorities.

MB FACT: Darkness reduces competition by limiting population growth. Fewer species survive, but those that do thrive. Resources stretch farther. Stability replaces abundance. Balance emerges.

MB FACT: Some species use bioluminescence to confuse predators by flashing behind them. The sudden light draws attention away from escape. Darkness favors misdirection. Confusion saves lives.

MB FACT: Sharks, though associated with lighted waters, rely heavily on non-visual senses. Electroreception and vibration detection work in darkness. Vision is only part of awareness. Familiar predators also think beyond sight.

MB FACT: Darkness allows extreme specialization. With fewer variables, evolution refines survival strategies. Niches become narrow but effective. Precision replaces versatility. Darkness sharpens design.

MB FACT: Some animals glow internally rather than externally. Light passes through translucent tissue. This creates soft illumination without sharp outlines. Darkness favors subtle glow. Visibility is controlled.

MB FACT: The *tripod fish* lives more than **3,000 feet deep** in the Atlantic and Pacific Oceans, where no sunlight reaches. It balances on long, stiff fin "legs" and faces into ocean currents, waiting for prey to bump into it. It doesn't chase food -- food comes to it.

MB FACT: The *barreleye fish*, found in the deep Pacific Ocean, has a **transparent head**. Its eyes sit inside a clear dome and rotate upward to detect faint light from above or silhouettes of prey. Bone would block too much light, so evolution removed it.

MB FACT: The *viperfish* lives between **2,000 and 9,000 feet deep** and hunts in total darkness. It has needle-like teeth and a glowing lure near its mouth. The light attracts prey close enough for a sudden strike -- no vision required.

MB FACT: The *vampire squid* inhabits oxygen-poor waters around **3,000 feet deep**, a zone deadly to most animals. Instead of chasing prey, it drifts slowly and feeds on falling organic debris. Its metabolism is so slow it can survive where others suffocate.

MB FACT: *Lanternfish*, found worldwide in deep oceans, use glowing light organs on their bellies to hide. By matching the faint light filtering down from above, they erase their shadows. Predators below see nothing at all.

MB FACT: The *giant squid* lives so deep and far from light that humans did not film one alive until **2004**. Its eyes are the largest on Earth, built to detect faint movement, not detail. In the dark, motion matters more than shape.

MB FACT: *Tube worms* at hydrothermal vents near mid-ocean ridges live **without mouths or stomachs**. Bacteria inside their bodies convert chemicals from Earth's interior into food. These animals thrive miles below the surface without ever seeing sunlight.

MB FACT: The *deep-sea dragonfish* produces red light -- something most deep-sea animals cannot see. This lets it illuminate prey without being detected. In the dark, it hunts with a private flashlight.

OCEAN PUNS
(that won't make you too salty)

**Keep your friends close...
and your anemones closer.**

What did the letter B say in the summertime?
"I do love to be beside the C."

Did you hear about her prom-ocean?

Sorry, can't work today.
I'm a little tide down at the moment.

There's no-fin better than a beach.

You don't like the sea?
Well... beach to their own.

**Buoy oh buoy,
what a wonderful day!**

Swim in that?
Are you squidding me?

**I've got a remedy for seasickness --
it's called a pocean.**

What is a scuba diver's favorite game?
Hide and sea-k.

**Why didn't the lobster share its food?
Because it was too shell-fish.**

How does seaweed answer the phone?
"Hello, how can I kelp you?"

MYTHS - BUSTED

MYTH: Darkness underwater means life must be rare or barely surviving. This idea assumes sunlight is required for all ecosystems. In reality, many dark-water environments are powered by chemistry, heat, and organic material sinking from above. Entire food webs operate without light. Darkness removes photosynthesis, not life.

MYTH: Animals that live in darkness are blind and poorly adapted. While some species lose eyes, many develop advanced non-visual senses like vibration detection, chemical sensing, and pressure awareness. These systems often outperform vision in dark environments. Losing eyes is not weakness -- it is efficiency. Darkness favors different strengths.

MYTH: Bioluminescence exists mainly to make animals visible to others. In most cases, glowing actually helps animals hide, hunt, or escape. Light can erase silhouettes, distract predators, or lure prey. Visibility is carefully controlled, not broadcast. In darkness, light is a tool, not decoration.

MYTH: Deep darkness makes underwater animals more aggressive because resources are scarce. Scarcity usually produces the opposite behavior. Many dark-water species conserve energy and avoid conflict whenever possible. Chasing or fighting wastes resources that may not be replaced for months. Calm behavior is often the winning strategy.

MYTH: Humans understand most environments that exist in permanent darkness. Large portions of the deep ocean, underwater caves, and deep lakes remain unexplored. Technology limits how long and how deep humans can observe. New species and behaviors continue to be discovered. Darkness still hides more than it reveals.

MYTH: If something glows underwater, it must be trying to attract attention. Many glowing animals produce light only when disturbed or threatened. Sudden flashes can confuse predators or draw attention away from escape. Glow does not always mean display. Sometimes it means defense.

This ancient **megalodon** lived between **23 million and 3.6 million years ago**, long before humans existed. It grew up to **60 feet long** -- about the length of **two school buses** -- and weighed as much as **60 tons**, heavier than a space shuttle. Its teeth could reach **7 inches tall**, larger than an adult human hand, with razor-sharp edges designed to slice through bone. Scientists estimate its bite force was the strongest of any animal ever known, powerful enough to crush whale skulls. When megalodon ruled the oceans, even other giant predators weren't safe.

REAL LEGENDS

LEGENDS

The Glow Beneath the Oars
Mediterranean Sea -- ancient coastal sailors

Sailors rowing along quiet coastlines on moonless nights described light moving beneath their boats, pacing them just below the surface. The glow was soft and steady, not flashing, and it followed the rhythm of the oars without ever breaking the water. Men slowed their strokes, afraid to disturb whatever traveled below.

No creature surfaced. No sound accompanied the light. When rowing stopped, the glow lingered, then slowly drifted away into deeper water. Elders warned crews not to row hard after dark, saying the sea sometimes "answered movement with movement." Even when explanations came later, the timing stayed unsettling.

The Cave That Blinked
Yucatán Peninsula -- early freshwater cave divers

Divers entering flooded caves reported flashes of light deep inside passages untouched by sunlight. The light appeared only when movement stirred the water, blinking faintly, then vanishing again. Some divers described the sensation that the cave was reacting to them.

When teams stopped moving, the light stopped too. Scientists later attributed the flashes to disturbed bioluminescent organisms clinging to cave walls. Still, divers remembered the silence afterward -- the way the cave returned to complete darkness, as if closing its eyes

---

The Black Escort,
Indian Ocean -- merchant sailing ships

Crews crossing calm seas at night recorded long, dark shapes pacing their vessels just beyond lantern glow. No wake broke the surface, and no outline could be defined. The movement was steady, deliberate, and silent. Sailors stood watch for hours as the presence followed, then gradually fell behind without warning. Logs described the event simply as "escorted passage." No one claimed danger -- only the feeling of being observed.

The Still Watchers of Baikal
Lake Baikal -- local fishermen and divers

Divers descending into Lake Baikal's deep, cold darkness reported shapes suspended far below, motionless and immense. These forms did not respond to light or sound. Nets set nearby often returned damaged without clear cause. Fishermen avoided those depths, believing size hid best when nothing moved. Scientists suggested optical distortion caused by depth and cold water layers. Locals still say Baikal holds its largest things perfectly still, waiting for darkness.

The Lights That Matched the Submersible
North Atlantic -- early deep-sea pilots

Pilots descending into deep darkness reported faint points of light appearing alongside their submersibles. The lights matched speed and depth precisely, neither approaching nor retreating. Each maneuver was mirrored, silently. When the submersible stopped, the lights stopped. When ascent began, the lights faded away. Later analysis suggested bioluminescent animals reacting to pressure and movement. But pilots recalled the precision most -- the feeling that something else understood exactly where they were.

On a Lighter Side

Fun Fact Comics

MIND-BLOWN™ Cartoons

DID YOU KNOW ?

Did You Know ? Humans rely on vision more than any other sense, which makes darkness feel threatening by default. Underwater, this bias becomes a liability because light disappears quickly and reference points vanish. Many early reports of strange behavior or size were shaped by this mismatch between human perception and underwater reality. When sight fails, the brain fills gaps with fear or imagination. Underwater life evolved without that bias. Darkness feels hostile to us, not to them.

Did You Know ? Permanent darkness allows ecosystems to remain stable for extremely long periods. Without sunlight-driven cycles, growth and change slow dramatically. This stability lets organisms specialize deeply rather than adapt constantly. Over time, these environments produce life forms that look alien simply because they evolved under different rules. Stability, not chaos, often defines the deep.

Did You Know ? Bioluminescence often works as camouflage rather than illumination. By matching faint background light or breaking up outlines, animals can erase their silhouettes completely. This means predators and prey may pass within inches of each other without detection. What looks bright in a lab can be nearly invisible in context. Light underwater is carefully measured, not freely given.

Did You Know ? Darkness forces animals to rely on physics instead of vision. Pressure changes, vibrations, and chemical trails become reliable information sources. These signals travel farther and more consistently than light. This allows animals to sense events happening well beyond visual range. Darkness reshapes intelligence by shifting which information matters most.

Did You Know ? Many dark-water legends began as honest observations with missing context. Sailors and divers often noticed movement, sound, or light without seeing a source. Without modern instruments, explanation lagged behind experience. Stories filled the gap between observation and understanding. Technology later confirmed that many of these events had real biological causes.

Did You Know ? Exploration of permanent darkness is still limited by human technology. Pressure, cold, and navigation challenges restrict how long we can observe deep environments. Most encounters are brief snapshots rather than long studies. This means behavior may be misunderstood or incomplete. Darkness continues to protect its secrets simply by being hard to access.

STORY MOMENT

STORY MOMENT

The Darkness That Answered Back

The diver checked his gauges one more time before pushing deeper into the flooded cave. Sunlight had already vanished behind him, replaced by a heavy blackness that swallowed the beam of his headlamp within a few yards. The water was perfectly still, thick and unmoving. Every movement, every fin kick, every breath -- felt louder than it should have, as if the cave were listening.

As he moved forward, the light swept across the rock wall.

Something flickered back.

He stopped instantly, hovering in place.

At first, he assumed it was a reflection, a trick of angle, a mineral surface catching the beam. He shifted position and swept the light again, slower this time. A second flash answered, dimmer and farther ahead in the passage. His breathing slowed. He waited, counting seconds.

The cave waited too.

When he moved, the light returned.

It wasn't random. Each shift of his fins produced a delayed response somewhere ahead -- never in the same spot, never forming a shape, never close enough to explain. When he backed away, the flashes faded. When he drifted forward again, they returned, scattered and quiet, like distant acknowledgments rather than reactions.

For several minutes, diver and darkness responded to each other without contact. No currents stirred. No animals crossed the beam. The water remained undisturbed except by the diver himself. And yet the timing felt wrong -- too measured, too patient, as if something was choosing when to reply.

Finally, training took over. He signaled his partner and began a slow, controlled retreat toward open water, careful not to rush. The flashes followed until the first hint of distant daylight appeared behind them. Then they stopped all at once. The cave fell completely dark again, as if nothing had ever happened.

Later, researchers suggested disturbed bioluminescent organisms along the cave walls. The explanation fit the basic data.

What it didn't explain was the pause.

The diver's log contained a single line:

"Light went in. Something answered. Darkness stayed."

FUN QUIZ

1. What powers most ecosystems in permanent underwater darkness?
   A) Sunlight reflected from the surface
   B) Photosynthesis by algae
   C) Chemistry, heat, and sinking organic material
   D) Tidal motion alone

2. Why do many deep-water animals lose color or become translucent?
   A) To scare predators
   B) Because pigment costs energy and offers no benefit in darkness
   C) Due to colder temperatures
   D) Because they live shorter lives

3. True or False: Most animals living in darkness are blind and poorly adapted.

4. What is one main purpose of bioluminescence in dark underwater environments?
   A) Decoration
   B) Warming the body
   C) Communication, camouflage, or hunting
   D) Navigation by stars

FUN QUIZ ANSWERS

1. C
2. B
3. False
4. C

CHAPTER 3

3-MOVING WATERS
Currents, Pressure, and Invisible Forces

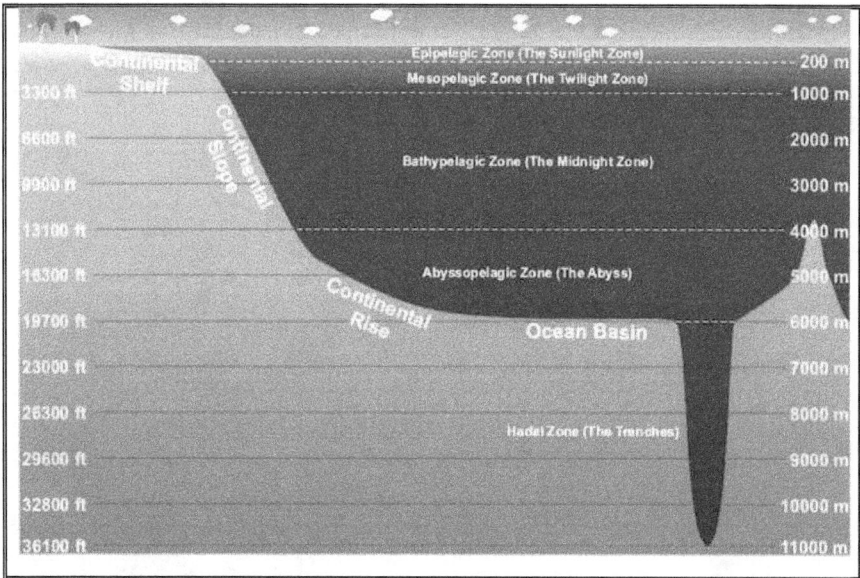

FUN & WEIRD FACTS

MB Fact = Mind-Blown Fact
Real discoveries. Real mysteries, Real Facts. Fun to make you think.

MB FACT: Water pressure increases rapidly with depth, rising by one atmosphere every 33 feet (10 meters). At depths where **sunlight disappears -- around 3,300 feet** -- pressure is already 100 times greater than at the surface. This extreme force crushes air-filled spaces, which is why **deep-sea animals have fluid-filled bodies instead of air**. Humans must rely on reinforced equipment to survive where marine life lives naturally.

MB FACT: The **barreleye fish** lives in the deep Pacific Ocean at depths of 2,000 to 2,600 feet, where only faint traces of light remain. Its **transparent skull** allows its eyes to rotate upward and capture every possible photon. Bone would block too much light, so **evolution removed it entirely**.

MB FACT: Internal waves travel underwater along layers of different temperature and density rather than on the surface. In the **South China Sea**, these waves can reach heights over **300 feet**, taller than many skyscrapers. Ships never notice them, but submarines feel sudden vertical movement when passing through invisible underwater hills.

MB FACT: The **SOFAR sound channel**, located roughly 3,000–4,000 feet deep, allows sound to travel **thousands of miles underwater**. Pressure and temperature bend sound waves inward instead of letting them fade. **Whale calls have crossed entire ocean basins** inside this natural sound tunnel.

MB FACT: The **vampire squid** lives in oxygen-minimum zones around 3,000 feet deep, where oxygen levels are too low for most fish. Instead of hunting, it feeds on **marine snow**, drifting organic debris. Its **extremely slow metabolism** allows survival where predators would suffocate.

MB FACT: More than **90% of animals below 1,000 meters** produce their own light through **bioluminescence**. Near the surface, fewer than 5% do. In darkness, light becomes a tool -- to lure prey, confuse predators, or **erase silhouettes**.

MB FACT: Hydrothermal vents form along mid-ocean ridges such as the **East Pacific Rise**, where Earth's crust is thin. Superheated water erupts from the seafloor, yet surrounding water stays near freezing. Entire ecosystems survive here because **Earth's internal heat replaces the Sun**.

MB FACT: The **giant squid** lives so deep that humans did not film one alive until 2004. Its eyes grow over **11 inches wide**, the largest on Earth. These eyes detect **movement, not detail**, which matters most in near-total darkness.

MB FACT: Deep-ocean circulation moves cold, dense water along the seafloor for thousands of miles in a global conveyor system. One full loop takes about **1,000 years**. This slow movement delivers oxygen and stabilizes Earth's climate.

MB FACT: Some underwater caves contain **haloclines**, where freshwater floats above saltwater. In the **Yucatán cenotes**, this boundary looks like underwater fog. When divers pass through, **vision blurs and bends** because light behaves differently in each layer.

MB FACT: Sound travels four times faster in water than in air, but unevenly. Temperature and pressure layers bend sound waves, making noises seem closer, farther away, or from the wrong direction. In the ocean, **hearing can lie**.

MB FACT: Many deep-sea fish sense Earth's magnetic field using **tiny magnetite crystals** embedded in their tissues. This allows navigation across open ocean with no landmarks. **Magnetism works where light cannot**.

MB FACT: Early life stages of many marine animals **drift instead of swim**. Fish larvae released near reefs may travel hundreds or thousands of miles. **Ocean currents decide destinations long before behavior does**.

MB FACT: Pressure changes how gases dissolve in water, including oxygen and carbon dioxide. At depth, gases stay dissolved more easily. When water rises quickly, bubbles can form -- **the same process behind decompression sickness**.

MB FACT: Some deep-sea animals use **counter-illumination** to hide in plain sight. Fish like **hatchetfish glow on their bellies** to match faint surface light. Predators below see no shadow -- **invisibility beats speed**.

MB FACT: Methane hydrates form under high pressure and low temperature along continental margins. Methane gas becomes trapped inside **ice-like crystals**. These deposits store **more carbon than all fossil fuels combined**.

MB FACT: Some underwater currents **reverse direction multiple times a day** due to tides interacting with seafloor shape. In submarine canyons, flow can flip within minutes. For local animals, **water motion becomes a schedule**.

MB FACT: Even in total darkness, the deep ocean is **never silent**. Earthquakes, shifting plates, and moving sediments send constant vibrations. Some animals detect these signals long before humans do -- **motion replaces sound**.

MB FACT: The **hadal zone**, deeper than 20,000 feet, is colder than Antarctica and under crushing pressure. Yet amphipods and snailfish live there using **special molecules that protect their proteins**. Without them, life would fail.

MB FACT: The **deepest-living fish ever recorded** was filmed in 2022 at 27,349 feet. Its bones are **thin and flexible**, preventing them from snapping. At extreme depth, **flexibility beats strength**.

MB FACT: The **Giant Phantom Jellyfish** can span over 30 feet and lives between 3,000 and 6,000 feet deep. It has been seen fewer than **150 times in history**. Its size and fragility make encounters extremely rare.

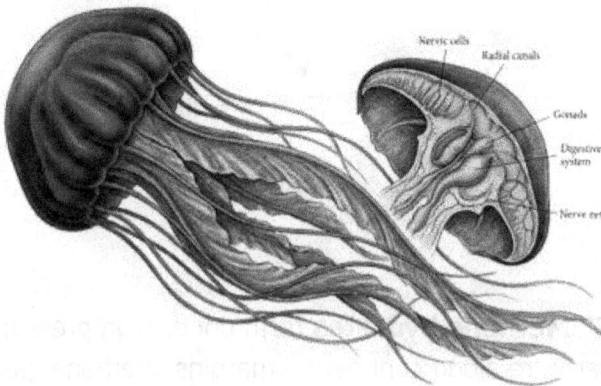

Giant Phantom Jellyfish
Stygiomedusa gigantea

MB FACT: Some deep-sea fish are **biologically locked to pressure**. When brought too close to the surface, their enzymes fail and cells shut down. For them, **surface water is lethal**.

MB FACT: Pressure at the bottom of the **Mariana Trench** can bend steel and crack glass. Submersibles use **curved titanium spheres** because curves distribute force evenly. **Geometry becomes survival**.

MB FACT: In the deep ocean, **red is the best camouflage color**. Red light disappears first underwater, so red animals appear black and invisible. What looks obvious to humans **vanishes completely at depth**.

MB FACT: The deep ocean floor preserves **human-made objects for centuries**. Glass bottles over 100 years old remain unchanged. Without sunlight or waves, **time barely touches the deep**.

MB FACT: Some deep-sea fish have **teeth so long they cannot close their mouths**. The **viperfish** uses needle-like teeth to grab rare prey in darkness, where **missing one meal can mean starvation**.

MB FACT: In total darkness, many deep-sea animals **do not follow day or night cycles**. Activity happens only when food arrives. **Opportunity replaces timekeeping**.

MB FACT: The **tripod fish** lives on the seafloor more than 3,000 feet down. It stands on long fin "legs" and lets currents deliver food. **Stillness saves energy** where swimming would waste calories.

Mariana Trench , 5800m + depth. ROV based imagery

ROV "Nereus" - Depth: 6,800m - Mariana Trench

LOCATION: CHALLENGER DEEP MARIANA TRENCH.
DEPTH: 10,890 M.

MYTHS - BUSTED

MYTH: If the ocean surface looks calm, the water below must be calm too. This belief comes from judging underwater conditions by sight alone. In reality, strong currents often move below a glassy surface, especially near cliffs, drop-offs, and shelves. Divers and instruments routinely measure powerful motion in visually calm seas. Surface calm does not equal underwater stillness.

MYTH: Underwater currents always move sideways like rivers. Many people imagine currents as horizontal flows only. In fact, water frequently moves upward or downward due to pressure, temperature, and density differences. These vertical currents can lift or sink objects without warning. The ocean moves in three dimensions, not one.

MYTH: Pressure underwater only affects how deep humans can go, not how water behaves. Pressure shapes how gases dissolve, how sound travels, and how buoyancy works. It changes how bodies move, think, and react. Animals evolved specifically to live inside this force. Pressure is not just a limit -- it is an active influence.

MYTH: If something pulls a diver or swimmer, it must be an animal. Sudden movement often triggers fear-based assumptions. In most cases, the force comes from changing currents, underwater slopes, or funneling passages. Water can accelerate around structures just like air around buildings. Physics explains the pull far more often than predators.

MYTH: Currents are chaotic and unpredictable. While they feel random to humans, many currents follow consistent patterns driven by tides, temperature, and seafloor shape. Animals learn and exploit these patterns. Scientists can often predict them with models and instruments. Unfamiliar does not mean unknowable.

MYTH: Only ocean environments experience strong underwater forces. Lakes, rivers, and flooded caves also contain powerful currents and pressure effects. Temperature differences and narrow passages create movement even in freshwater. Some inland waters are more dangerous because they appear harmless. Water physics does not depend on salt.

DEEP SEA CONFUSION!

MIND-BLOWN™ Cartoons

REAL LEGENDS

The Day the Sea Pulled Sideways
North Atlantic -- trawler crews near the continental shelf

Fishermen working a calm morning reported nets drifting sharply to port without wind or swell. The boat's compass held steady, but lines angled hard as if caught by a moving wall. Engines strained to compensate, yet the pull persisted -- quiet, constant, and wrong.

When the nets finally surfaced, they were twisted and scoured, as though dragged across rock that charts insisted wasn't there. Old hands warned the crew to clear the edge before the "sideways water" tightened. Scientists later explained accelerated flow along the shelf break, where depth drops suddenly. The crews still remember the feeling: not being pulled forward or back, but **taken sideways by the sea itself**.

The Pressure That Stole Time
Red Sea -- early commercial divers

A diver descending along a reef wall felt the world slow. Tasks that should have taken seconds stretched oddly, as if thought itself thickened. Hand signals came late; responses lagged. The water pressed in, steady and patient.

When he surfaced, the team compared logs and found gaps -- minutes missing where none should be. Doctors later named nitrogen narcosis and pressure effects on cognition. The diver never argued the science. He only said the sea didn't rush him. **It waited, and time bent with the depth.**

The Current That Chose a Direction
Pacific Islands -- lagoon fishermen

At slack tide, canoes drifted without aim inside a sheltered lagoon. Then, without warning, every floating leaf turned and slid the same way. Paddles bit water, but progress slowed as if the lagoon had decided on a route of its own.

Elders said the water "found its door." Oceanographers later pointed to density differences and a narrow channel opening offshore. The fishermen still watch for the moment when everything points at once. **When water agrees, resistance becomes pointless.**

The Cave That Breathed
Mediterranean coast -- freedivers

Divers entering a flooded sea cave felt a gentle surge outward, then inward, like a slow inhale. There was no wave at the entrance, no wind to explain it. Lights revealed silt lifting, settling, then lifting again.

Tide charts showed nothing unusual. Later studies suggested pressure pulses and internal tides pushing water through the cave's narrow throat. The divers accepted the model -- and kept the name. **They said the cave breathed, and they timed their dives to its lungs.**

On a Lighter Side

THE CAVE THAT BREATHED

DID YOU KNOW ?

Did You Know ? Water can move powerfully without forming visible waves. This happens because underwater motion is often driven by density, temperature, and pressure differences rather than wind. A calm surface can hide strong movement below, especially near shelves, cliffs, or narrow passages. Humans tend to trust what they can see, which makes this especially misleading. Instruments regularly detect fast-moving water beneath glassy seas. Stillness at the surface is not proof of safety below.

Did You Know ? Pressure underwater affects the human brain as well as the body. At depth, gases dissolve into tissues more easily, which can subtly alter thinking and reaction time. Divers may feel calm and confident while actually being impaired. This makes decision-making harder, not easier. The danger isn't panic -- it's false clarity. Pressure can quietly change how reality feels.

Did You Know ? Underwater sound does not travel in straight lines. Pressure layers bend sound waves, causing them to curve or bounce unexpectedly. This can make noises seem closer, farther, or even come from the wrong direction. Animals evolved to interpret these distortions naturally. Humans often misjudge distance and location because of them. Hearing underwater is shaped by physics, not intuition.

Did You Know ? Some underwater currents move on daily or seasonal schedules. Temperature changes between day and night can shift water density enough to start motion. These predictable flows deliver food and oxygen to certain areas at specific times. Animals time feeding and movement around these cycles. To them, currents act like clocks. Timekeeping underwater often depends on motion, not light.

Did You Know ? Underwater structures -- both natural and human-made -- reshape water flow dramatically. Rocks, reefs, wrecks, and caves force water to speed up, slow down, or change direction. These altered flows create shelter zones where animals gather. They also create hazards where water accelerates unexpectedly. The shape of the environment determines how water behaves.

Did You Know ? Large-scale underwater currents help regulate Earth's climate. They transport heat from warm regions to colder ones and redistribute nutrients around the globe. Without these slow, massive movements, surface temperatures would shift dramatically. Weather patterns on land depend on motion far below the surface. The ocean quietly stabilizes the planet through movement. Invisible water shapes visible life.

On a Lighter Side

DEEP SEA CONFUSION!

MIND-BLOWN™ Cartoons

STORY MOMENT

STORY MOMENT

Pulled Without a Hand

The diver had done this route dozens of times. A slow descent along the wall, a short survey at depth, then back up before the current shifted. The sea looked perfect that morning -- flat, quiet, cooperative. Nothing suggested trouble.

At twenty meters, he felt it.

Not a yank. Not a jerk. Just a steady sideways pressure, like leaning against a moving train. He adjusted his fins and kicked gently, expecting the resistance to fade. It didn't. The wall slid past him faster than it should have, even though he wasn't swimming.

He checked his depth. Stable. His breathing was calm. Still, the pull increased.

There was no sound, no swirl of sand, no visible motion in the water. The sensation was unsettling precisely because it was smooth. If something had grabbed him, panic would have made sense. Instead, it felt like the water had chosen a direction and included him in it.

He angled upward, thinking elevation might help. The pull softened slightly, then returned stronger. His computer showed nothing unusual. The sea did not appear dangerous. That made it worse.

Training cut through the confusion. He stopped fighting and changed strategy -- turning his body to let the current slide past instead of through him. Slowly, the pressure eased. The wall stopped moving. The water let go.

Later, on deck, researchers explained the cause: a fast-moving boundary current flowing along the drop-off, invisible from above. A thin layer of water moving faster than the one he entered.

The explanation fit perfectly.

What stayed with him wasn't fear, but respect. The realization that water doesn't need to rush or roar to be powerful.

Sometimes, it just moves -- and expects you to move with it.

"PULLED WITHOUT A HAND"

MIND-BLOWN™ Cartoons

FUN QUIZ

1. Why can underwater currents be strong even when the surface looks calm?
   A) Surface waves cancel them out
   B) Currents are driven by density, temperature, and pressure differences
   C) Wind hides their motion
   D) Only animals create underwater movement

2. Which type of current moves water up or down instead of sideways?
   A) Coastal currents
   B) Boundary currents
   C) Vertical currents
   D) Tidal wakes

3. True or False: If a diver feels pulled underwater, it is usually caused by an animal.

4. Why do some animals conserve energy by positioning themselves in currents?
   A) Currents provide warmth
   B) Currents reduce pressure
   C) Currents allow movement without muscle effort
   D) Currents confuse predators

5. How does pressure affect human thinking underwater?

   A) It increases alertness

   B) It has no effect on the brain

   C) It can slow judgment and reaction time

   D) It improves decision-making

FUN QUIZ ANSWERS

1. B

2. C

3. False

4. C

5. C

4-LOST TO THE ABYSS

Shipwrecks, Sunken Cities, and Strange Things Left Behind

FUN & WEIRD FACTS

MB Fact = Mind-Blown Fact
Real discoveries. Real mysteries, Real Facts. Fun to make you think.

MB FACT: In **1968**, a Soviet submarine called **K-129** sank in the Pacific Ocean more than **16,000 feet deep**. The wreck was so secret that the U.S. government tried to lift part of it using a ship disguised as a deep-sea mining vessel. The mission was called **Project Azorian** and stayed classified for decades. One sunken submarine triggered one of the strangest spy operations in history.

MB FACT: The ocean floor near **Scapa Flow, Scotland**, is packed with sunken German warships from **World War I**. In **1919**, German crews deliberately sank their own fleet to keep it from being captured. Today, massive battleships lie upside down like toppled toys. It's one of the largest intentional ship graveyards on Earth.

MB FACT: The wreck of the **SS Thistlegorm**, sunk in **1941** in the Red Sea, is filled with motorcycles, trucks, boots, and even train parts. These supplies were meant for British troops but never made it ashore. Divers can still see vehicles lined up exactly where they were stored. It's like opening a supply crate frozen in time.

MB FACT: In **Lake Huron**, a shipwreck called the **Cornelia B. Windiate** sits upright on the lake floor after sinking in **1875**. The ship went down so gently that its dishes are still stacked neatly in cupboards. Cold freshwater helped keep everything in place. Sometimes sinking doesn't mean smashing.

MB FACT: The **Baltic Sea** is so good at preserving wrecks that even ropes and leather shoes survive. A medieval shipwreck discovered near **Sweden** still had coiled rope on deck after **700**

years underwater. Low salt and cold temperatures stopped bacteria from doing their job. The sea accidentally became a time capsule.

MB FACT: During World War II, hundreds of airplanes crashed or landed in shallow tropical waters. In places like **Palau** and **Chuuk Lagoon**, planes now rest underwater with coral growing over wings and cockpits. Fish swim through where pilots once sat. Nature reclaimed the machines without erasing them.

MB FACT: Some shipwrecks weren't lost to storms or war -- they were sunk on purpose to block enemies. During **World War II**, ships were intentionally sunk to block harbors and rivers. These wrecks acted like underwater roadblocks. Ships became barriers instead of vehicles.

MB FACT: Ancient sailors often threw cargo overboard to survive storms. Amphora jars, anchors, and tools still litter the seafloor along ancient trade routes in the **Mediterranean Sea**. These underwater trails show where ships traveled thousands of years ago. The ocean keeps receipts.

MB FACT: In **2019**, scientists found a shipwreck in the **Arctic Ocean** that was almost perfectly preserved in icy water. The cold slowed decay so much that wooden beams still looked freshly cut. Ice didn't crush the ship -- it protected it. Cold can save history.

MB FACT: Long ago, the Roman emperor **Caligula** built two giant party ships on **Lake Nemi**. These weren't normal boats -- they had marble floors, statues, plumbing, and even heated rooms. Yes, on a lake. The ships sank nearly **2,000 years ago**, turning a quiet lake into a Roman mystery.

MB FACT: Lake Baikal is so deep and cold that it hides shipwrecks like a freezer hides food. In the early **1900s**, ferries carried train cars across the frozen lake during winter. Some ships sank and are still down there today. A lake once helped run a railroad.

MB FACT: Divers in **Lake Van** didn't expect to find a castle underwater -- but they did. Stone walls from a **3,000-year-old fortress** sit beneath the lake's surface. The water slowly rose over time and swallowed it whole. The lake didn't crash in -- it crept.

MB FACT: Some shipwrecks act like underwater weather makers. As water hits the wreck, it creates swirling currents that trap food particles. Fish learn these spots quickly and gather there. A broken ship can turn into the busiest place on the seafloor.

MB FACT: When the passenger ship **RMS Titanic** sank in **1912**, it didn't reach the seafloor intact. At about **12,500 feet deep**, pressure crushed internal air spaces and caused the ship to split in half *before* impact. The bow and stern landed more than **2,000 feet apart**, with a debris field covering **over 15 square miles**. That's why explorers didn't find "the ship" in one place -- only a trail of destruction.

MB FACT: In the **Black Sea**, shipwrecks can survive almost perfectly preserved. Below about **600 feet**, the water contains almost no oxygen, which stops wood-eating organisms from living there. Ancient Greek ships dating back to **400 BCE** have been found upright, with masts still standing. Oxygen -- not age -- is what usually destroys ships.

MB FACT: The warship **USS Oriskany** was deliberately sunk in **2006** off the coast of **Florida** to create an artificial reef. The ship weighs over **30,000 tons** and sits in about **210 feet of water**. Within a single year, it became home to thousands of fish and corals. Structure creates life fast when the ocean needs shelter.

MB FACT: Many shipwrecks are found far from where ships were last reported. When Air France Flight **447** crashed into the Atlantic in **2009**, debris sank through multiple current layers moving in different directions. The wreckage was eventually found **nearly 13,000 feet deep**, spread over a wide area. Water keeps moving long after gravity pulls things down.

MB FACT: Some wooden shipwrecks survive longer than steel ones. In the **Great Lakes**, cold freshwater slows corrosion and prevents salt damage. Ships from the **1800s** still show paint, railings, and nameplates. Salt -- not time -- is the real destroyer.

MB FACT: Steel wrecks decay from the inside out due to electrochemical corrosion. The **Titanic's** hull is being eaten by bacteria that form rust structures nicknamed **"rusticles."** Scientists estimate the wreck could collapse completely within the next **30–50 years**. Even giants dissolve when chemistry takes over.

MB FACT: Some shipwrecks move long after sinking. The German battleship **Bismarck**, sunk in **1941**, rests on a sloping seafloor nearly **15,000 feet deep**. Over time, parts of the wreck have slid downhill as metal weakened. The seafloor is not flat -- or stable.

MB FACT: Entire ancient cities now lie underwater due to rising sea levels. Near **Atlit Yam** in Israel, stone houses and wells dating back **9,000 years** sit beneath the Mediterranean. These settlements weren't destroyed by war or disaster. The ocean simply rose and never went back down.

MB FACT: Submarines are among the hardest wrecks to find. The U.S. submarine **USS Grayback**, lost in **1944**, was not located until **2019**. A single mistranslated word in wartime records placed it **100 miles** from its real location. Paper errors can hide steel giants.

MB FACT: Some shipwrecks vanish without being destroyed. In strong-current areas like the **English Channel**, sand can bury entire wrecks, then uncover them decades later. A wreck marked on maps can simply disappear -- and reappear -- without moving. The ocean hides things by covering them, not breaking them.

MB FACT: The Swedish warship **Vasa** sank on its very first voyage in **1628**, less than **1 mile** from shore in Stockholm Harbor. It went down because the ship was top-heavy, with too many cannons stacked high and not enough ballast below. Cold, low-salinity Baltic Sea water preserved the wooden hull almost perfectly. That's why a 400-year-old ship can still stand upright today.

MB FACT: In **1985**, explorers **Robert Ballard** and his team finally located the **Titanic** using robotic cameras, not submarines. They followed a trail of coal and debris across the seafloor instead of searching for the ship itself. This technique changed how wrecks are found worldwide. Sometimes the mess tells the story better than the object.

MB FACT: Lake Issyk-Kul may be hiding entire lost towns. Stone walls, tools, and ancient graves have been found underwater near old trade routes. Earthquakes and rising water likely pulled the cities down. One of Asia's busiest travel routes might now be underwater.

MB FACT: In **Lake Fuxian**, sonar scans revealed massive stone platforms and pyramid-like shapes on the lake floor. Some are longer than a football field. No stories or records explain who built them. The lake kept the secret.

MB FACT: The **River Thames** isn't just water -- it's a history drawer. Ancient boats over **4,000 years old** have been found buried in its muddy bottom. The mud blocks oxygen, stopping wood from rotting. The river quietly saved them.

MB FACT: The ancient city of **Heracleion**, once a major Egyptian port, sank beneath the Mediterranean around **1,200 years ago**. Earthquakes and liquefaction caused the land to collapse, allowing seawater to rush in. Statues weighing **over 5 tons** now lie scattered underwater. The city didn't erode -- it dropped.

MB FACT: The warship **USS Arizona**, sunk during the **1941 Pearl Harbor attack**, still leaks small amounts of oil today. This slow seepage is known as "**the tears of the Arizona**." The ship rests in shallow water, but its steel tanks still hold fuel decades later. Sunken ships can remain chemically active for generations.

MB FACT

The warship USS Arizona, sunk during the 1941 Pearl Harbor attack, still leaks small amounts of oil today.

This slow seepage is known as "the tears of the Arizona". The ship rests in shallow water, but its steel tanks still hold fuel decades later. Sunken ships can remain chemically active for generations.

MB FACT: The **Antikythera wreck**, discovered in **1900** off Greece, contained the world's oldest known analog computer. The device dates back to **around 100 BCE** and was used to predict eclipses and planetary movement. It was so advanced that scientists didn't understand it fully until the **2000s**. Ancient technology can rival modern expectations.

MB FACT: In **2018**, a perfectly preserved **World War II aircraft carrier** -- the **USS Lexington** -- was found **10,000 feet deep** in the Coral Sea. The ship still has aircraft inside with visible paint and markings. Cold, deep water slowed decay dramatically. Depth can protect as much as it destroys.

MB FACT: Some shipwrecks create "ghost fishing" zones. Lost fishing nets and lines remain attached to wrecks and continue trapping fish for **decades**. This happens because modern plastics do not break down easily. A wreck can keep killing long after humans leave.

MB FACT: The **Yonaguni Monument** near Japan lies about **80 feet underwater** and features massive stone steps and terraces. Some scientists believe it is a natural formation; others argue it shows signs of human shaping. If man-made, it would date back over **10,000 years**, before known civilizations in the region. The debate continues because the ocean erased the context.

MB FACT: During World War II, over **7,500 ships** were sunk globally, many never mapped. Modern sonar surveys still discover wrecks every year in busy shipping lanes. Some are found beneath modern ports and oil platforms. History didn't sink quietly -- it stacked up.

MB FACT: Shipwrecks can change local currents. Large wrecks act like underwater hills, forcing water to rise, slow, or swirl around them. This creates feeding zones where fish gather in higher numbers than surrounding areas. A single wreck can reshape an ecosystem without moving an inch.

MB FACT: The ship **Mary Rose**, flagship of King Henry VIII's navy, sank in **1545** just off the coast of **England** during a battle. It tipped over suddenly after firing its cannons, letting water rush in through open gun ports. When it was raised in **1982**, scientists found shoes, tools, weapons, and even dice still inside. The wreck became a frozen moment of Tudor life.

MB FACT: In **Lake Superior**, shipwrecks sit in water so cold it acts like a refrigerator. The freighter **SS Edmund Fitzgerald**, which sank in **1975**, rests in **530 feet** of icy freshwater. Because there's no salt and very little oxygen, the wreck has barely changed in decades. Cold water slows decay more than depth alone.

MB FACT: The ancient city of **Pavlopetri** off the coast of **Greece** lies just **10–13 feet underwater** and dates back over **5,000 years**. Streets, courtyards, and buildings are still clearly visible. The city wasn't destroyed by war -- it was slowly flooded as sea levels rose. Sometimes history disappears quietly, not violently.

MB FACT: The German submarine **U-576**, sunk in **1942** off **North Carolina**, was discovered in **2014** resting just **240 yards** from the cargo ship it attacked. Both wrecks sit on the seafloor like a frozen battle scene. This happened because they sank almost simultaneously during the same fight. The ocean preserved a moment of combat.

MB FACT: The wreck of the **San José**, a Spanish treasure ship sunk in **1708**, was found off **Colombia** in **2015**. It carried gold, silver, and emeralds worth billions today. The ship was lost during a naval battle, not a storm. Treasure ships didn't just sink by accident -- they were targets.

MB FACT: In the **Baltic Sea**, dozens of Viking ships lie underwater in near-perfect condition. Low salt levels and cold temperatures prevent shipworms from surviving there. Some wrecks still show carved wooden details from over **1,000 years ago**. Geography decides what history survives.

MB FACT: During World War II, the island of **Bikini Atoll** became an underwater ship graveyard. Nuclear bomb tests in **1946** sank battleships, cruisers, and submarines in the lagoon.

Today, those wrecks sit in clear tropical water, slowly growing coral. Weapons testing accidentally created one of the world's strangest wreck collections.

MB FACT: The ancient lighthouse city of **Alexandria, Egypt**, partially collapsed into the sea after earthquakes around **1,300 years ago**. Giant stone blocks and statues now lie underwater near the harbor. These weren't dumped -- they fell straight down when the ground failed. Earthquakes can erase cities without warning.

MB FACT: Some shipwrecks are discovered by accident during cable or pipeline surveys. Sonar scans meant for engineering often reveal outlines of ships no one knew were there. This is how several **World War I** wrecks were found in the **North Sea**. History hides in plain sight.

MB FACT: In **1943**, a German submarine sank a cargo ship off the coast of **Florida**, just **15 miles from the beach**. People standing on shore could actually see the ship burning before it went down. The wreck still sits offshore today, reminding historians that World War II reached U.S. waters. Wars don't always happen far away.

MB FACT: The **Great Lakes** hold more than **6,000 shipwrecks**, many perfectly preserved. Cold freshwater and low salt slow rust and rot so much that ships from the **1800s** still look recognizable. Some even have intact wheels and railings. The lakes quietly keep history instead of destroying it.

MB FACT: In **2011**, scientists discovered a shipwreck in the **Gulf of Mexico** covered in what looked like giant icicles. They weren't ice -- they were bacteria eating oil leaking from the wreck. The microbes created rock-hard tubes as they fed. Even bacteria get busy around shipwrecks.

MB FACT: In the **Rhône River**, a Roman cargo ship sank near **Arles** nearly **2,000 years ago**. The river covered it fast, protecting pottery, tools, and wood. It's like the river pressed "pause" on Roman life.

MB FACT: Around **Lake Constance**, scientists found ancient wooden homes built on stilts along the shore. These houses are over **5,000 years old**. When the water rose, the homes didn't break -- they flooded. Villages slipped underwater instead of vanishing.

MB FACT: **Lake Champlain** hides real warships from the **American Revolution** and the **War of 1812**. Cold freshwater kept cannons and decks intact. Some wrecks still look battle-ready. History didn't sink far -- it sank quietly.

MB FACT: At the bottom of **Lake Tahoe**, old steamships rest where tourists once traveled in the **1800s**. When newer boats arrived, some were simply sunk instead of removed. Cold, deep water slowed decay. Even vacation lakes have secrets.

MB FACT: The ancient city of **Dwarka** off the coast of **India** lies underwater and is mentioned in texts over **2,000 years old**. Stone walls, anchors, and streets have been found beneath the sea. Rising ocean levels likely swallowed the city slowly. Legends sometimes sink instead of disappear.

MB FACT: Some shipwrecks still have working lights -- sort of. In rare cases, bioluminescent organisms gather inside portholes and corridors. When disturbed, the wreck briefly glows blue. It's not electricity -- it's living light.

MB FACT: The wreck of the **USS Indianapolis**, sunk in **1945**, wasn't found until **2017**. It lay more than **18,000 feet deep**, far deeper than most wrecks. Survivors of the sinking faced days in the water, making it one of the most tragic naval disasters in U.S. history. Finding the wreck finally closed a 72-year mystery.

MB FACT: Some ancient anchors found underwater weigh over **10 tons** and are carved from solid stone. These were used by early civilizations before metal anchors existed. Losing one meant losing the ship. Anchors were once more valuable than cargo.

MB FACT: Shipwrecks can sound creepy even when nothing is there. Water moving through broken hulls creates moaning and knocking noises. Divers have reported hearing "footsteps" caused entirely by currents. The ship isn't haunted -- physics is just loud underwater.

MB FACT: During World War II, pilots sometimes ditched planes at sea instead of crashing. Some of these aircraft landed gently enough that they sank almost intact. Today, explorers find fighter planes sitting upright on the seafloor like they just parked there. The ocean can be surprisingly gentle.

MB FACT: In **Japan**, fishermen have found shipwrecks covered in glass bottles that still hold liquid from over **100 years ago**. With no sunlight and little movement, the bottles stayed sealed and unchanged. The deep ocean can act like a locked drawer. Some things stay exactly where they fell.

MB FACT: In the **Caribbean**, pirate-era shipwrecks still hide coins, cannons, and personal items. Many sank during storms while fleeing naval patrols in the **1600s and 1700s**. Treasure wasn't always buried on islands -- sometimes it never made it out of the hold. Storms ended pirate careers fast.

MB FACT: Off the coast of **Colombia**, the Spanish treasure ship **San José** sank in **1708** during a cannon battle. It carried gold, silver, and emeralds from South America meant for Spain's king. The wreck wasn't found until **2015**, resting more than **2,000 feet deep**. One ship became one of the most valuable shipwrecks ever discovered.

MB FACT: Near the Greek island of **Antikythera**, divers found an ancient shipwreck in **1900** filled with statues, jewelry, and a strange set of bronze gears. The device, now called the **Antikythera Mechanism**, was built around **100 BCE**. Scientists later discovered it could predict eclipses and planetary motion. An ancient ship was carrying a mechanical brain.

MB FACT: In the waters off **India**, stone walls and streets have been discovered beneath the sea near **Dwarka**, a city mentioned in texts more than **2,000 years old**. Sonar scans and dives revealed anchors, buildings, and paved areas. Rising sea levels likely swallowed the city slowly. Legends don't always vanish -- they sink.

MB FACT: The ancient Egyptian city of **Heracleion** disappeared under the Mediterranean around **1,200 years ago**. Earthquakes caused the ground to liquefy, letting massive stone temples slide into the sea. Divers later found statues weighing **over 5 tons** lying underwater. The city didn't erode -- it collapsed.

MB FACT: In **Lake Titicaca** between Peru and Bolivia, archaeologists have found submerged stone structures at high altitude. These ruins lie beneath freshwater, not ocean water. The lake once had lower levels, and ancient people built along its shores. When the water rose, the land -- and history -- went under.

MB FACT: Ancient sailors often hid treasure inside ceramic jars called **amphorae**. Hundreds have been found in shipwrecks across the **Mediterranean Sea**, some still sealed after **2,000 years**. These jars once held oil, wine, coins, or spices. The seafloor became a storage vault.

MB FACT: Near **Japan's Yonaguni Island**, massive underwater stone steps and platforms sit about **80 feet below the surface**. Some researchers think they are natural rock formations, while others believe humans shaped them over **10,000 years ago**. Sharp angles and straight lines keep the mystery alive. The ocean blurred the evidence.

MB FACT: Roman ships carried heavy stone anchors carved from single blocks. Many of these anchors now sit underwater near ancient ports in **Italy and Greece**. Some weigh more than **10 tons** and still show rope grooves. Losing one could strand an entire fleet.

MB FACT: Archaeologists have found ancient trade ships off the coast of **Turkey** carrying cargo from three different continents. These wrecks date back more than **3,000 years** and held copper, glass, and luxury goods. They prove that global trade existed long before modern maps. The sea connected the ancient world.

On a Lighter Side

In hindsight, letting the octopus
handle navigation was optimistic.

MIND-BLOWN™
Cartoons

MYTHS - BUSTED

MYTH: Pirates cursed their buried treasure.

Pirates across the Caribbean believed treasure needed protection after burial. Legends say captains performed rituals or left a victim behind to "guard" the gold. Anyone digging in the wrong place -- or at the wrong time -- would be struck by illness or madness. Treasure wasn't just hidden; it was claimed.

MYTH: The Flying Dutchman appears before disaster.

Sailors from Europe to South Africa feared sightings of the ghost ship. Seeing its glowing sails was said to guarantee shipwreck or death before the next sunrise. Even seasoned captains recorded sightings in logs. The myth survived centuries because crews swore it came true.

MYTH: The Black Sea preserves ships because it is cursed.

Ancient sailors believed something unnatural about the Black Sea kept wrecks intact. Ships didn't rot, bodies didn't decay, and wood stayed whole. Long before science discovered the anoxic layers, sailors said the sea refused to let things disappear.

MYTH: Lake Baikal swallows ships whole.

Local legends in Siberia warned that Lake Baikal "opens" during storms. Boats vanished without debris or bodies, as if pulled straight down. Fishermen said the lake chose when to give things back. Even today, some losses remain unexplained.

MYTH: Mermaids warned sailors of danger.

Early myths didn't paint mermaids as villains. Sailors believed their songs were warnings about rocks, storms, or deadly currents. Ignore the song, and the sea would take you anyway. Hearing one meant you were already close to trouble.

MYTH: The Bermuda Triangle consumes ships on purpose.

For centuries, sailors avoided the region between Florida, Bermuda, and Puerto Rico. Compasses failed, storms formed suddenly, and ships vanished without distress calls. Before modern explanations, crews believed the ocean itself was hostile there.

MYTH: Whirlpools were living mouths.

Norse sailors feared giant whirlpools like the Maelstrom. Ships weren't pulled in by water -- they were swallowed. Survivors claimed the sea chewed vessels apart before spitting out wreckage days later. Calm water afterward meant the creature was full.

MYTH: Some wrecks are guarded by their crews.

Divers and fishermen reported shadows, snagged nets, and sudden equipment failures near certain wrecks. Sailors believed the last crew stayed behind to protect the ship. A guarded wreck was one you didn't touch.

MYTH: Rivers remember what they take.

River communities believed lost boats, weapons, or bodies would resurface years later. Objects appeared miles downstream long after sinking. The river decided when to give things back. Until then, it kept them.

MYTH: Pirates chained treasure to the sea floor.

Not all treasure was buried on land. Some legends say pirates dropped chests overboard and chained them to reefs or rocks. Only captains knew the exact spot. Miss it, and the treasure stayed lost forever.

MYTH: The Great Lakes keep their dead.

Sailors feared cold lakes like Lake Superior because bodies often never floated back up. The water was said to "hold on" to what it claimed. The phrase "the lake never gives up her dead" became more than a saying.

MYTH: Ghost lights mark shipwrecks.

Sailors reported blue or green lights hovering above wreck sites. These lights were believed to be trapped souls or warnings to stay away. Seeing one meant danger nearby.

MYTH: Some ships sank twice.

Old stories tell of wrecks that vanished, reappeared years later as hazards, then disappeared again. Storms uncovered them briefly before sand buried them deeper. The sea rearranged its memories.

MYTH: Underwater cities drowned silently.

Legends from around the world claimed cities didn't fall in disasters -- they were abandoned quietly as water rose. Doors closed, streets flooded, and people left without panic. The sea didn't destroy them; it replaced them.

MYTH: The Devil's Sea erased ships near Japan.

East of Japan, sailors feared the Devil's Sea, where ships vanished and navigation failed. Area earned a reputation similar to the Bermuda Triangle. Even government ships avoided it.

MYTH: Some wrecks curse salvagers.

Stories tell of storms, accidents, and deaths following attempts to recover artifacts. Crews believed certain ships did not want to be disturbed. Leave them alone -- or pay the price.

MYTH: Sea monsters followed ships, not the other way around.

Sailors believed massive creatures tracked vessels for days, just below the surface. Shadows moved with the ship, never attacking, just watching. Being followed was worse than being attacked.

REAL LEGENDS

LEGENDS

The Harbor That Slipped Away
Eastern Mediterranean -- ancient coastal settlements

Fishermen along the eastern Mediterranean passed down stories of stone roads that once led straight into the sea. Nets snagged on squared blocks far offshore, and anchors scraped across worked stone instead of sand. Elders said the shoreline had not always been where it was now.

Over generations, the water crept forward, swallowing docks, walls, and homes without a single night of disaster. Archaeologists later confirmed entire harbor districts now rest underwater, preserved beneath layers of sediment. The legend wasn't about destruction. It was about how quietly a city can disappear when the sea decides to keep moving.

The Ship That Sank Too Slowly
North Sea -- 19th-century merchant routes

Crews spoke of a cargo ship that didn't plunge when it was damaged but seemed to settle gradually, deck by deck. Lanterns stayed lit longer than expected as water crept upward. Sailors swore the ship gave them time -- too much time -- to think.

Later investigation suggested trapped air and compartmented holds slowed the descent. Survivors remembered the stillness most. The sea did not rush the ship. It waited until gravity finished the conversation.

The Lake That Gave Back Streets

Central Europe -- drought years

Villagers recalled summers when stone walls and roads emerged from shrinking reservoirs. Old maps matched what appeared: staircases, foundations, even doorways. Children played where boats usually passed.

When the rains returned, the streets vanished again. Historians confirmed entire towns had been submerged during dam construction decades earlier. The legend became a reminder that water does not erase history -- it stores it.

The Convoy That Never Arrived

Pacific Ocean -- wartime supply routes

A wartime convoy vanished between island chains without distress calls. Years later, fishermen reported nets tearing on metal far below known shipping lanes. Sonar eventually revealed multiple wrecks scattered across miles of seabed.

Currents and storms had pulled sinking ships apart mid-descent, spreading them far from their last positions. The legend of disappearance gave way to a harder truth: nothing vanished. Everything fell, just not together.

The Flying Dutchman

Ancient Sea Monster of Myths and Legends

DID YOU KNOW ?

Did You Know ? Shipwrecks often become easier to find decades after they sink, not immediately. As metal corrodes or wood weakens, parts collapse and spread outward, increasing the detectable footprint. Currents may also remove sand that once buried the wreck. This means a wreck can seem to "appear" long after the event. Discovery timing often has more to do with environmental change than technology. The sea reveals things gradually.

Did You Know ? Rising sea levels have submerged thousands of former coastlines worldwide. During the last ice age, oceans were much lower, exposing land where people lived, traded, and built. As glaciers melted, water advanced inland slowly, not suddenly. Many communities relocated without catastrophe. Their structures remained behind, waiting underwater. Loss did not always feel dramatic at the time.

Did You Know ? Underwater decay depends heavily on oxygen availability. In oxygen-poor water, bacteria and organisms that normally break down wood and metal cannot survive. This allows ships, docks, and even rope to persist far longer than expected. Some environments essentially pause decomposition. Preservation is chemical, not accidental. Darkness and low oxygen protect history.

Did You Know ? Human-made objects change underwater ecosystems in lasting ways. A single wreck can redirect currents, trap sediment, and concentrate food sources. Over time, this reshapes where animals live and hunt. Removing a wreck may disrupt an established balance. What began as debris often becomes infrastructure. The sea adapts faster than people expect.

Did You Know ? Not all sunken cities are ancient. Some were submerged deliberately during modern dam projects. Entire towns were evacuated, documented, and then flooded. When water levels drop, buildings sometimes reappear almost intact. These places feel ancient but are not. History can be surprisingly recent.

Did You Know ? Many underwater discoveries happen by accident rather than design. Fishing nets, anchors, and construction surveys often uncover wrecks unintentionally. Even advanced sonar surveys miss objects hidden by terrain or sediment. Chance still plays a major role in exploration. The ocean does not reveal everything on demand. It shows what it chooses, when conditions allow.

STORY MOMENT

The Last Thing They Saw Was the Clock

The ship wasn't supposed to sink. That's what everyone agreed on later.

It was a short crossing, clear weather, routine cargo. When the impact came, it felt more like a mistake than a disaster -- a shudder through the hull, a pause, then confusion. Water began entering low and slow, filling spaces no one watched closely enough.

By the time the captain ordered abandon ship, the deck already tilted. Crew members moved carefully, stepping around loose gear, passing familiar objects they suddenly realized they would never see again. Someone tried to save a toolbox. Someone else stopped to grab a photograph. Both were told to leave it.

As lifeboats pulled away, the ship didn't plunge. It settled.

Lights stayed on longer than expected. Windows dipped below the surface one by one. The last visible thing wasn't the mast or the bridge -- it was the clock near the stern, still ticking as water climbed past it.

Then the sea closed.

Years later, a survey vessel detected the wreck far from its last reported position. Currents had carried it sideways during descent, laying it gently on a slope. Divers found plates stacked in the galley, boots lined by bunks, the clock still mounted but silent.

Nothing looked violent. That was what unsettled them.

The ship hadn't been destroyed. It had been placed.

The ocean hadn't erased the crossing. It had archived it -- every decision, every moment of hesitation, every object left behind because there wasn't time.

The wreck now sits quietly, reshaping currents, sheltering fish, collecting silt. Above it, ships still pass without knowing.

Below, time stopped exactly where the water reached the clock.

So where's everyone else?

FUN QUIZ

---

1. Why are many shipwrecks found far from where they were last seen on the surface?
   A) Ships often change direction while sinking
   B) Currents can move wreckage sideways during descent
   C) Maps were inaccurate in the past
   D) Ships are usually dragged by sea animals

2. What factor most strongly determines whether a wreck is well preserved underwater?
   A) The size of the ship
   B) How famous the ship was
   C) Water chemistry and oxygen levels
   D) The speed at which it sank

3. True or False: Most sunken cities were destroyed suddenly by massive disasters.

4. Why can shipwrecks collapse or shift decades after they sink?
   A) Earthquakes happen constantly
   B) Gravity, corrosion, and currents continue to weaken them
   C) Marine animals pull them apart
   D) Salvage crews trigger movement

5. Why do shipwrecks often attract large numbers of fish and marine life?
   A) Wrecks generate heat
   B) Wrecks block sunlight
   C) Wrecks create shelter and change local water flow
   D) Fish are curious about metal

FUN QUIZ ANSWERS

1. B

2. C

3. False

4. B

5. C

CHAPTER 5

5-SIGNALS FROM DEEP
Weird Sounds, Strange Echoes, and Underwater Mysteries

FUN & WEIRD FACTS

MB Fact = Mind-Blown Fact
Real discoveries. Real mysteries, Real Facts. Fun to make you think.

MB FACT: In **Loch Ness**, near Inverness in northern Scotland, sound can slide sideways through deep cold layers instead of rising. The lake plunges over **750 feet deep**, and noise made far below can travel unseen and unheard along the bottom. To someone on shore, the water can seem silent while signals move underneath.

MB FACT: Beneath the **Pacific Ocean** west of California and around Hawaii lies the **SOFAR channel** *(short for Sound Fixing and Ranging channel)*, a natural sound tunnel about **half a mile down**. In this layer, sound barely weakens. Scientists once sent a signal near Australia that was detected near the U.S. West Coast -- **over 3,000 miles away**.

MB FACT: In **1997**, underwater microphones across the **South Pacific Ocean** recorded a massive sound later called **The Bloop**. It was loud enough to reach sensors thousands of miles apart. For years, no animal fit the signal's size -- until it was traced to enormous Antarctic ice cracking underwater.

MB FACT: During the Cold War, listening stations in the **North Atlantic** between the United States and the United Kingdom tracked submarines by their engine hums alone. Every submarine produced a slightly different sound pattern, called an **acoustic signature** *(a unique noise fingerprint)*. Crews could identify a vessel without ever seeing it.

MB FACT: Under the winter ice of **Lake Baikal** in Siberia, sound reflects like it's trapped in a tunnel. The lake is over **one mile deep**, and explorers have reported hearing deep knocks and long groans traveling beneath the frozen surface. The ice turns the lake into a giant echo chamber.

MB FACT: Near the **Mid-Atlantic Ridge**, underwater sensors regularly pick up low rumbles from the seafloor. These signals come from shifting rock and volcanic movement, not animals. The ocean quietly listens to the Earth reshaping itself far below.

MB FACT: In the **Arctic Ocean** north of Alaska, cracking sea ice produces sharp pops and long moans that can travel huge distances underwater. These sounds are strongest during seasonal freeze and thaw. To early sailors, the ice sounded alive.

MB FACT: Along the **Great Barrier Reef**, underwater microphones record nonstop crackling from snapping shrimp. Each snap creates a tiny shockwave that can briefly exceed **200 decibels**. Millions snapping together turn the reef into a constant wall of sound.

MB FACT: Off the coast of **California**, submarines rely on passive listening instead of sonar pings. Engines, propellers, and movement leave sound trails that travel far underwater. Staying silent is safer than being invisible.

MB FACT: In deep freshwater lakes like **Lake Superior**, sound can become trapped below cold surface layers in early spring. Noise spreads sideways instead of upward, creating hidden activity zones. Calm water on top doesn't mean quiet water below.

MB FACT: In the **Mediterranean Sea** near Italy, modern cargo ships create low-frequency sounds that travel dozens of miles underwater. These signals are so steady that scientists can tell when shipping traffic increases -- just by listening. The sea keeps a record of human movement.

MB FACT: In the **Gulf of Mexico**, underwater sensors have recorded repeating low-frequency sounds linked to distant storms. The signals arrive **hours before** weather reaches land. The water hears the storm before the sky shows it.

MB FACT: Off the coast of **Newfoundland** in the North Atlantic, scientists once turned the ocean into a giant "thermometer" using **acoustic thermometry** *(measuring ocean temperature by timing sound travel)*. Because sound speed changes with temperature, tiny warming shifts show up as measurable timing changes across long distances. In other words: the ocean can be "read" by how fast a signal arrives.

MB FACT: Near **Diego Garcia** in the **Indian Ocean**, scientists use the **SOFAR channel** *(short for Sound Fixing and Ranging channel)* to listen across entire ocean regions. Low-frequency sounds entering this layer can travel **thousands of miles** with little loss. One signal can pass beneath ships, storms, and darkness without fading.

MB FACT: In the **Strait of Gibraltar**, sound behaves unpredictably as Atlantic water and Mediterranean water collide. Different temperatures and salinity bend signals sharply. A noise made near Spain can arrive from the direction of Africa, confusing both animals and instruments.

MB FACT: Deep in the **Mariana Trench**, underwater microphones have recorded low rumbles caused by pressure shifts and distant earthquakes. Some of these signals travel upward from nearly **36,000 feet down**. Even the deepest place on Earth sends messages to the surface.

MB FACT: In **Lake Tanganyika**, one of the deepest freshwater lakes in the world, sound travels far along steep underwater walls. Fishermen have reported hearing boat noises long before seeing anything. The lake acts like a natural sound corridor.

MB FACT: Beneath the **North Sea**, sonar surveys often detect echoes that seem larger than they really are. Dense schools of fish and old shipwrecks can merge into one massive signal. Early sonar operators sometimes mistook these stacked echoes for giant objects.

MB FACT: In the **Black Sea**, thick layers of water trap sound at certain depths. Signals can travel sideways for long distances without rising. This creates hidden acoustic zones where movement is heard but never seen.

MB FACT: Along the **Puerto Rico Trench**, underwater sensors record deep cracking sounds linked to tectonic stress. Some signals repeat without warning. The ocean here listens to the planet flexing beneath it.

MB FACT: In winter on **Lake Geneva**, temperature layers form that bend sound sharply. A splash on one side of the lake can echo strangely on the other. Calm water doesn't mean predictable sound.

MB FACT: In the **Baltic Sea**, low oxygen and cold water preserve old shipwrecks that still produce noise. Expanding metal and shifting sediment create slow creaks picked up by listening devices. Wrecks continue to signal long after sinking.

MB FACT: Off the coast of **Monterey Bay**, scientists discovered that underwater landslides make deep, rolling sound waves long before sediment reaches the bottom. These signals can travel for miles through the bay. The ocean often *hears* a collapse before anyone can see it.

MB FACT: Near **Juan de Fuca Ridge**, hydrothermal vents release superheated water that makes constant low bubbling and roaring sounds. These signals helped scientists locate vents before cameras ever saw them. Heat leaves fingerprints you can hear.

MB FACT: In the **Mediterranean Sea** near Italy, researchers discovered that methane gas escaping the seafloor creates rhythmic popping sounds. Each bubble release sends a tiny acoustic pulse upward. The seabed quietly "breathes" through sound.

MB FACT: In the **Persian Gulf**, shallow water causes ship noise to reflect repeatedly between the surface and seafloor. This stacking effect makes engines sound louder and closer than they really are. In tight seas, sound exaggerates reality.

MB FACT: In **Chesapeake Bay**, **male toadfish** produce a steady underwater hum during mating season. Entire sections of the bay vibrate for hours at a time. Locals once thought boats were idling underwater all night.

MB FACT: Beneath the **Southern Ocean**, underwater microphones have detected sounds from icebergs scraping the seafloor as they drift. These signals travel far and change pitch as the iceberg rotates. Even moving ice leaves an audio trail.

MB FACT: Off **Hawaii**, scientists use underwater microphones to detect whale calls bouncing off underwater mountains. The returning echoes reveal the shape of the seafloor. Whales accidentally help humans map the deep.

MB FACT: In **Lake Erie**, strong winds can cause a hidden water movement called a seiche *(a slow sloshing of the entire lake)*. As the water shifts, underwater sensors pick up long, rolling pressure sounds. The lake signals its own motion.

MB FACT: Near **Guam**, underwater sensors have detected the sound of meteors hitting the ocean surface. The impact sends a sharp acoustic spike through the water. Space can announce itself through the sea.

MB FACT: Along the **Norwegian Sea**, schools of herring produce coordinated clicking sounds when threatened. The noise confuses predators by flooding the water with overlapping signals. Safety comes from sounding bigger than you are.

MB FACT: During the Cold War, listening stations in the **North Atlantic** detected unknown repeating sounds that turned out to be drifting icebergs far offshore. The signals arrived days before the ice was ever seen. Sound became early warning.

MB FACT: In **Lake Nyos**, Cameroon, underwater sensors detected strange pressure sounds before a deadly gas release in **1986**. Carbon dioxide had been building up silently at the bottom, and tiny acoustic changes were the only warning. The lake signaled danger before anyone knew how to listen.

MB FACT: During World War II, underwater microphones near **Scotland** picked up unusual clicking sounds that turned out to be enemy mines activating. The sounds didn't explode immediately -- they warned first. Listening saved ships before radar ever could.

MB FACT: Off the coast of **Oregon**, USA, underwater microphones record sharp cracking sounds caused by sand dunes moving along the seafloor. As currents push sediment downhill, grains collide and vibrate. Even sand sends signals when it moves.

MB FACT: In the **Red Sea**, near the coasts of Egypt and Saudi Arabia, dolphins use rapid burst-clicks to warn others when sharks are nearby. These signals are short, intense, and directional. It's an underwater alarm system built from sound alone.

MB FACT: Beneath **Lake Tahoe**, on the border of California and Nevada, USA, deep monitoring equipment has recorded long, low vibrations caused by underwater slope shifts. These signals can happen without earthquakes or storms. The lake quietly adjusts itself -- and announces it through sound.

MB FACT: In the **Bering Sea**, between the United States and Russia, drifting sea ice collides underwater, producing deep booming tones instead of sharp cracks. These low sounds can travel far below the surface. Ice doesn't just break -- it signals.

MB FACT: Near **Santorini**, Greece, underwater microphones monitor volcanic activity by listening for pressure pops and rumbling pulses. Sound changes often appear before eruptions. The volcano speaks before it erupts.

MB FACT: In **Hudson Bay**, Canada, beluga whales use unique call patterns that bounce off ice edges to locate openings for air. The echoes tell them where breathing holes are forming. Sound becomes survival.

MB FACT: Along the **Yellow Sea**, bordering China and South Korea, underwater noise from strong tides creates rhythmic pressure signals that repeat like a heartbeat. Sensors can identify tide strength by sound alone. The sea keeps time acoustically.

MB FACT: In deep sections of the **Caribbean Sea**, near Puerto Rico and the Lesser Antilles, collapsing coral structures create brief popping sounds as trapped air escapes. These signals alert nearby fish to sudden habitat changes. Reefs announce damage in real time.

MB FACT: Off the coast of **Iceland**, underwater microphones record harmonic humming sounds caused by ocean waves interacting with submerged lava formations. The shape of the rock controls the pitch. The seafloor can hum like an instrument.

MB FACT: Off **Svalbard**, Norway, underwater microphones record long, flute-like tones when strong ocean swells pass over shallow seabed ridges. The waves squeeze water through narrow gaps, turning rock shapes into sound makers. The seafloor can whistle when conditions line up.

MB FACT: In **Lake Titicaca**, shared by Peru and Bolivia, pressure sensors detect low vibrations when cold nighttime air cools the surface quickly. The rapid temperature change makes the water contract and shift, sending signals through the lake. Even cooling water can announce itself.

MB FACT: Near **Papua New Guinea**, underwater recorders have captured sharp clicking bursts from reef fish coordinating night movements. The clicks are brief and directional, helping groups stay together in darkness. Sound becomes a flashlight when light is gone.

MB FACT: Off **Cape Verde** in the eastern Atlantic, drifting pumice from underwater eruptions collides and scrapes, creating steady rasping sounds. These signals can last for days as the stones spread. Floating rock can make noise miles from the eruption.

MB FACT: In **Lake Rotorua**, geothermal bubbles rise from the bottom and pop in repeating patterns. Each burst sends a tiny acoustic pulse upward. The lake quietly broadcasts heat from below.

MB FACT: Along the coast of **Japan**, underwater sensors pick up rapid pressure pulses when powerful typhoons pass offshore. The signals arrive underwater before winds hit land. The sea warns first.

MB FACT: In the **Gulf of California**, Mexico, snapping fish produce synchronized clicks during spawning events. The sound spikes help scientists pinpoint breeding grounds without seeing a single fish. Signals reveal gatherings hidden from view.

MB FACT: Beneath **Lake Kivu**, Rwanda and the Democratic Republic of the Congo, sensors record faint tremor-like sounds as gas layers shift at depth. These signals change with seasons and pressure. The lake's chemistry speaks quietly.

MB FACT: Off **Scotland** near the Outer Hebrides, underwater microphones capture low humming created when strong tides rush through narrow channels. The water itself vibrates as it accelerates. Fast flow turns movement into sound.

MB FACT: In the **Coral Sea**, Australia, nighttime recordings show reefs getting louder after sunset as animals switch to sound-based signaling. The rise in noise marks a daily shift from sight to vibration. Darkness flips the signal system on.

MB FACT: In the **GIUK Gap**, NATO listening stations tracked submarines by the tiny vibrations from their propellers. Even when engines were quiet, the spinning blades made rhythmic pressure pulses. The ocean turned mechanics into signals.

MB FACT: Off **Norfolk**, the world's largest naval base area, underwater microphones can hear ships long before they appear. Hull shape and speed change how water vibrates around a vessel. Ships announce themselves just by moving.

MB FACT: In the **South China Sea**, sonar operators deal with a problem called **reverberation** *(echoes bouncing repeatedly in* *shallow water)*. One sound can return many times from different directions. The sea can multiply a single signal into confusion.

MB FACT: During **World War II, convoy routes in the Arctic Ocean** were monitored by listening for ice-crushing ship hulls. As vessels forced through pack ice, the pressure created long grinding sounds underwater. Even reinforced ships couldn't hide their struggle.

MB FACT: Near **Pearl Harbor**, modern sensors can identify boats by **cavitation noise** *(tiny bubbles collapsing near propellers)*. Faster movement creates louder bubble pops. Speed leaves a sound trail.

MB FACT: In the **English Channel**, dense shipping traffic fills the water with overlapping engine tones. Listening stations can tell busy days from quiet ones just by sound levels. Human activity hums constantly below the waves.

On a Lighter Side

If we had one wish, we'd still use it for different music.

If you listen closely you can hear the sounds of traffic.

MYTHS - BUSTED

MYTH: Whistling at sea could summon storms and disaster.

For centuries, sailors believed that whistling on a ship was dangerous. They thought the sound mocked the wind and angered powerful sea spirits, calling up violent storms in response. On some ships, whistling could earn punishment, or worse...especially during calm weather, when sailors feared tempting the ocean's mood. To whistle was to challenge forces no one could control.

MYTH: Selkies were seals that could become human by shedding their skin.

In Scottish and Irish coastal legends, selkies slipped out of their seal skins and walked ashore as humans under the moonlight. If someone stole the skin, the selkie was trapped on land, often forced into marriage and a life away from the sea. But the stories never ended happily, because the call of the ocean was stronger than love. Once the skin was found, the selkie always returned to the water, leaving everything behind.

MYTH: Magical islands appeared and vanished in the open ocean.

Sailors in the North Atlantic told stories of islands like **Hy-Brasil** that could suddenly rise from the mist and then disappear again for years. Some claimed the islands were covered in gold, eternal sunlight, or immortal beings. Others said they only appeared once every seven years- and vanished the moment someone tried to land. Maps showed them, then erased them, then showed them again, as if the ocean itself couldn't decide what was real.

MYTH: The ocean was born from the blood of a slain giant.

In Norse mythology, the gods Odin, Vili, and Ve killed the giant **Ymir** and used his body to create the world. His flesh became the land, his bones the mountains, and his skull the sky. His blood poured into the endless void called the "Abyss of Abysses," flooding it and forming the sea. The ocean was not gentle - it was violent, ancient, and born from destruction.

REAL LEGENDS

LEGENDS

The Humming Sea
Pacific Northwest -- coastal fishing grounds

Fishermen working night lines began reporting a low, steady hum rising from beneath their boats. It wasn't wind, and it wasn't the engine. The sound pulsed slowly, strong enough to vibrate hull planks and make loose gear rattle. Crews cut engines and listened as the water itself seemed to resonate.

Some said the hum followed them when they drifted. Others claimed it faded when they moved closer to shore. The noise never broke the surface, never revealed a source. Older fishermen warned that the sea was "working something out" and advised leaving the grounds alone until morning.

Years later, researchers linked similar sounds to spawning toadfish and resonance created by shallow coastal basins. The explanation fit the data, but not the feeling. Those who heard it still describe the moment when the water stopped being silent -- and started answering back.

The Echo That Wouldn't Match
Mediterranean Sea -- naval sonar crews

A patrol vessel recorded a sonar return that made no sense. The echo was too large, too slow, and too close to the hull to match any known object. Operators recalibrated instruments, then shut them down entirely. When sonar resumed, the signal was still there.

The echo mirrored the ship's movement, holding distance without approaching or retreating. It faded only when the vessel crossed a sharp temperature boundary. Analysts later determined layered water had bent sound waves into a false reflection.

The crew accepted the report. What stayed with them was the precision -- the way the echo behaved as if it understood spacing. For hours, something seemed to listen as closely as it was being listened to.

The Singing Ice
Lake Baikal -- winter crossings

Travelers crossing frozen Lake Baikal described eerie sounds rising from below the ice. Groans, whistles, and deep vibrations rolled across the lake, sometimes loud enough to wake sleepers. The noises felt alive, as if the ice itself was breathing.

Locals warned newcomers not to panic. The sounds intensified during temperature shifts, especially at night. Scientists later explained the phenomenon as pressure cracks and vibration traveling through ice and water together.

Still, those who heard it remembered the timing -- how the lake waited until darkness, then sang. Knowledge explained the cause. It didn't erase the feeling of being listened to by something vast and patient beneath the ice.

The Call Beneath the Fog
North Atlantic -- lighthouse keepers

Lighthouse logs recorded strange calls drifting in with thick fog. The sounds were low and rhythmic, carrying far beyond where visibility ended. Keepers described them as voices that didn't belong to wind or surf.

Some believed ships were signaling in distress. Others feared something moving unseen beneath the fog line. Investigations later pointed to whale calls bending through temperature layers and fog-heavy air.

What the logs show is hesitation. Lights stayed lit longer. Horns were sounded "just in case." The fog hid the source, but the sound reached shore anyway, reminding listeners that the sea speaks farther than it shows.

The Lake That Spoke at Dusk
Central European reservoir -- post-flood settlements

After an old town was submerged to create a reservoir, residents began hearing low knocks and distant tones at dusk. The sounds echoed faintly across the water, strongest near where streets once ran. Some said it was the town settling. Others said it was the lake remembering.

Engineers eventually traced the sounds to pressure changes moving water through submerged structures. Air pockets collapsed slowly. Walls redirected vibrations. The drowned town became an instrument.

People accepted the explanation, but the habit remained. At dusk, locals still pause near the shore -- listening for a place that no longer exists, except in sound.

On a Lighter Side

For a modest fee, you can now get ocean sounds on your smartwatch.

DID YOU KNOW ?

Did you know? Sound behaves so strangely underwater that it can feel **intelligent**. As sound bends around **temperature layers**, echoes can shift, return late, or seem to react to movement. Early sonar operators reported the eerie feeling of something "watching back." Nothing alive was there -- but physics was reshaping the signal in real time, creating the illusion of intent.

Did you know? Scientists often **hear underwater events before they ever see them**. Acoustic sensors can detect **landslides, ice breaks, and distant storms** long before cameras or satellites confirm anything happened. Sound travels faster and farther than light underwater, flipping how humans normally understand cause and effect. In the deep, sound arrives first -- answers come later.

Did you know? Some **freshwater lakes are harder to understand acoustically than oceans**. Steep underwater walls and sharp temperature layers can trap sound, bounce it sideways, or send it back multiple times. A single noise can echo, overlap, and distort itself. Calm-looking lakes can hide chaotic soundscapes below the surface.

Did you know? Many aquatic animals don't hear sound the way humans do -- they **feel it**. Vibrations move straight through water into **skin, bone, and tissue**, allowing detection of motion without seeing anything at all. What humans treat as background noise becomes critical information. Underwater, vibration replaces vision.

Did you know? Human activity has permanently altered the ocean's **soundscape**. Ships, sonar, and offshore construction add constant low-frequency noise that never fully fades. Some animals adapt by changing their calls or timing. Others leave entire regions behind. The sea didn't slowly grow louder -- it was made that way.

Did you know? Some underwater sounds remain unexplained simply because they are **rare**, not dramatic. A signal that happens once every few years may never repeat in the same way. Without patterns, confirmation becomes difficult. Science relies on repetition -- and one-time sounds can stay mysterious for decades.

Did you know? The ocean can hide sound in invisible **shadow zones**. At certain depths, sound bends away instead of spreading out, leaving areas where signals simply vanish. Something loud can pass nearby without being heard at all. Silence underwater doesn't always mean distance -- it can mean concealment.

Did you know? Some underwater echoes return **out of order**. Sound can reflect off multiple surfaces and arrive back at sensors seconds apart, even though it came from a single source. Early listeners sometimes thought several objects were moving when there was only one. The deep can multiply a moment into confusion.

Did you know? Night changes how water carries sound. As surface water cools after sunset, sound paths can shift, allowing signals to travel farther than during the day. This makes nighttime underwater listening more sensitive -- and more misleading. Darkness doesn't just hide movement; it reshapes how signals travel.

STORY MOMENT

The Signal That Wouldn't Fade

The hydrophone had been lowered hundreds of meters into the water, far below where light mattered and color disappeared. On deck, the lake looked calm -- gray, flat, unremarkable. Wind barely touched the surface. That was what made the sound so unsettling when it slipped into the headphones, quiet but unmistakable.

At first it was barely there. A low pulse, slow and steady, like a distant engine that never arrived. Too even. Too patient. The technician adjusted the gain, expecting it to dissolve into background noise. It didn't. The signal sharpened instead, repeating at precise intervals, bending slightly as the boat drifted and the cable swayed beneath them.

They cut the engine. The lake went still.
The sound remained.

Charts showed nothing unusual beneath them -- no wrecks, no steep drop-offs, no known structures. The bottom was smooth, mapped, familiar. Yet the pulse seemed to follow, always holding the same distance, as if aware of movement rather than reacting to it. Someone joked about turning the hydrophone *toward* the sound, then stopped laughing when the signal shifted again, subtly but clearly.

For nearly an hour, they listened. Each time the boat changed heading, the sound curved with it, arriving late, then early, then strong again. Sometimes it seemed to hesitate, as if deciding where to reappear. No pattern matched a machine. No behavior fit an animal. It wasn't loud -- just persistent, impossible to ignore once noticed.

When the temperature data finally came back, the explanation appeared. A sharp thermal layer sat beneath the surface, trapping sound and bending it inward. Signals were being folded back on themselves, looped and redirected by invisible walls of warm and cold water. The lake had become a mirror -- one that reflected sound instead of light.

They documented the data, logged the readings, and powered up. As the hydrophone rose through the layers, the signal thinned, stretched, and unraveled, losing its shape before fading completely.

On shore, everything was silent again. Wind in the grass. Waves at the rocks. Nothing out of place.

But long after the report was filed, the technician admitted something hadn't left him. For a while, the lake hadn't just carried sound.

It had seemed to listen.

FUN QUIZ

1. Why can underwater sounds sometimes seem to "follow" a moving boat?

A. The sound source is actively tracking the vessel

B. Water pressure amplifies sounds near metal objects

C. Sound is bent and redirected by temperature layers

D. Engines create echoes that repeat automatically

2. Why can scientists detect underwater events before seeing any physical evidence?

A. Light travels farther underwater than sound

B. Sound moves faster and farther underwater than light

C. Underwater cameras record continuously

D. Water magnifies all vibrations instantly

3. Which situation is most likely to create misleading sonar readings?

A. Calm water with no temperature differences

B. Shallow water with flat sandy bottoms

C. Schools of fish and uneven underwater terrain

D. Clear water with no suspended particles

4. Why can deep freshwater lakes be acoustically confusing?

A. Freshwater blocks sound more than saltwater

B. Lakes are too small for sound to reflect

C. Temperature layers and steep walls distort sound paths

D. Sound only travels vertically in lakes

5. Why do many aquatic animals detect movement earlier than humans underwater?

A. They have larger ears

B. They rely on vision instead of sound

C. They sense vibration through their bodies

D. Water amplifies sound only for animals

6. What is a "sound shadow zone" underwater?

A. An area where sound becomes louder due to pressure

B. A place where sound bends away and becomes hard to detect

C. A region blocked by solid rock or ice

D. A zone created only by human-made noise

7. Why do some underwater sounds remain unexplained for long periods of time?

A. Scientists avoid studying rare signals

B. The sounds are always too faint to measure

C. The events happen rarely and don't repeat

D. Equipment cannot survive deep water

8. Why can sound behave unpredictably in lakes as well as oceans?

A. Lakes are too shallow for sound to travel far

B. Freshwater reflects sound more than saltwater

C. Temperature layers and underwater shapes affect sound paths

D. Wind on the surface controls all underwater sound

FUN QUIZ ANSWERS

1. **C** -- Sound bends and curves through temperature layers, making it seem responsive
2. **B** -- Sound travels faster and farther than light underwater
3. **C** -- Dense fish groups and uneven terrain can stack or distort echoes
4. **C** -- Temperature layers and steep walls cause sound to bounce unpredictably
5. **C** -- Many animals feel vibration through tissue, not just ears
6. **B** -- Sound bends away in certain conditions, creating areas where signals seem to vanish
7. **C** -- Rare, one-time signals are difficult to confirm without repetition
8. **C** -- Temperature layers and underwater features bend and reflect sound

CHAPTER 6

6-FRESHWATER CAN GET WEIRD

The Strangest Life in Lakes and Rivers

FUN & WEIRD FACTS

MB Fact = Mind-Blown Fact
Real discoveries. Real mysteries, Real Facts. Fun to make you think.

MB FACT: In **Lake Erie**, freshwater can flip from safe to deadly almost overnight. Temperature layers collapse, oxygen drops, and chemical signals change suddenly, forcing fish to rely on rapid sensing instead of routine. Survival here depends on detecting change early, not being perfectly adapted. Freshwater life listens constantly.

MB FACT: In muddy backwaters of the **Mississippi River**, largemouth bass hunt using vibration instead of sight. Pressure waves from moving prey are detected by the **lateral line** *(a vibration-sensing system along the body)*. Even in zero visibility, motion leaves a readable signal. Darkness doesn't hide movement.

MB FACT: Along the **Ohio River**, catfish "taste" the water around them. Taste receptors across their skin and whiskers detect chemical signals released by food and other animals. In brown, lightless water, chemistry replaces vision. The river speaks in flavor.

MB FACT: Some animals don't hear danger through water at all -- they feel it through mud. **Crayfish** detect low-frequency vibrations traveling through sediment, letting them sense footsteps, falling rocks, or predators before anything enters the water. In streams of the Ozark Plateau, the riverbed itself acts like an alarm system. The ground warns first.

MB FACT: Losing eyesight isn't a failure -- it's an upgrade. **Cave-dwelling fish** gradually abandon vision and redirect energy to pressure and chemical sensing instead. In underground waters of the Dinaric Alps, movement, flow, and scent replace light completely. Darkness reshapes biology.

MB FACT: Some fish know a better habitat is nearby before they ever reach it. **Snakeheads** detect subtle chemical differences between connected waters, guiding them toward favorable conditions. In floodplains of the Yangtze River Basin, sensing comes first -- movement follows. Exploration starts invisibly.

MB FACT: During late summer in **Lake Winnipeg**, oxygen levels can crash suddenly as algae die off. Fish sense this change chemically and behaviorally, fleeing before conditions turn lethal. Those that fail to detect it in time suffocate. Survival depends on reading invisible signals fast.

MB FACT: In swampy waters of **Okefenokee Swamp**, bowfin survive where oxygen is scarce by detecting surface air and gulping it. Their modified swim bladder acts like a lung, letting them respond to chemical stress signals that kill other fish. Ancient design thrives in modern chaos.

MB FACT: After floods in the **Amazon River**, fish must relearn their environment using sound, vibration, and chemical traces. Channels shift, landmarks vanish, and familiar signals disappear. Memory and sensory flexibility matter more than instinct. Rivers erase maps regularly.

MB FACT: In northern lakes of **Canada**, northern pike rely on pressure signals to time ambush strikes. Instead of chasing prey, they wait for the exact vibration pattern that signals vulnerability. One explosive movement ends the hunt. Timing is read through water movement.

MB FACT: Some fish survive freezing winters by almost shutting down. Under the ice of **Lake Superior**, certain species reduce movement so much that only faint pressure changes reveal they're alive. Chemical cues tell them when it's safe to stay still and when to move again. Survival sometimes means becoming nearly invisible.

MB FACT: Northern pike don't chase -- they wait. In weedy shallows of **Lake Simcoe**, they read vibration patterns to decide the exact moment to strike. One explosive burst decides everything. Freshwater predators win by timing, not endurance.

MB FACT: Some fish navigate without ever seeing where they're going. In dark stretches of the **Danube River**, migration follows chemical gradients rather than landmarks. These signals persist even when channels change. Water carries directions long after maps fail.

MB FACT: Beneath winter ice on **Lake Baikal**, some freshwater fish survive by reducing activity almost to zero. Chemical and pressure cues tell them when movement is too costly. Stillness becomes a response to environmental signals. Doing nothing can be the correct move.

MB FACT: In turbid waters of **Lake of the Woods**, walleye hunt using low-light vision combined with vibration detection. Their eyes amplify available light while their bodies read motion through water pressure. Multiple signals merge into advantage. Murk favors specialists.

MB FACT: Sudden danger doesn't always look dramatic -- it smells wrong. Many freshwater fish respond to rapid chemical changes that signal collapsing oxygen or rising stress. In shallow prairie lakes of North Dakota, entire populations relocate within hours. The warning arrives before the crisis.

MB FACT: Group awareness can move faster than thought. **Perch** schools transmit vibration signals so quickly that a threat sensed by one fish spreads through hundreds almost instantly. In clear lakes near Lake Geneva, the group reacts as a single organism. Safety scales with numbers.

MB FACT: Some fish populations are older than the lakes they live in. As glaciers retreated and waterways shifted across Scandinavia, chemical cues guided survivors into new basins.

Water bodies changed, but signal-following behavior endured. Continuity hides under disruption.

MB FACT: Survival skills can become ecological weapons. **Tilapia** constantly adjust behavior in response to temperature, oxygen, and chemical signals, allowing them to dominate unstable waters. In reservoirs across East Africa, adaptability turns into takeover. Flexibility has consequences.

MB FACT: In the shallow bays of **Lake Champlain**, **common carp** use vibration and pressure cues to navigate murky water. Studies show they can remember feeding locations for **months**, responding to familiar movement patterns rather than sight. After being caught once, many carp avoid similar vibrations entirely. Sound and motion become memory.

MB FACT: In underground streams of **Mammoth Cave**, cave-dwelling fish survive in complete darkness by abandoning vision altogether. Over generations, they lose eyes and pigmentation while strengthening vibration and chemical sensing. Water movement replaces sight as the primary signal. Darkness forces sensory re-engineering.

MB FACT: Along the **St. Lawrence River**, **lake sturgeon** detect spawning grounds using subtle current changes and chemical traces. Some individuals live over **100 years**, relying on long-term stability of these signals. When flow patterns change, entire generations are affected. Longevity depends on reliable information.

MB FACT: Gar look prehistoric -- and they act like it. In backwaters of the **Trinity River**, they rise to gulp air when chemical signals warn of low oxygen. Their swim bladder functions like a lung. Ancient design still solves modern problems.

MB FACT: Some freshwater fish experience the night as a completely different world. In **Lake Okeechobee**, vibration patterns change after sunset as predators and prey swap roles. Signals that meant danger by day can mean opportunity by night. Darkness rewrites the rules.

MB FACT: Lamprey don't bite -- they attach. In rivers feeding **Lake Huron**, they locate hosts by detecting pressure changes and chemical trails. Their jawless mouths reflect a body plan older than dinosaurs. Freshwater still hosts ancient strategies.

MB FACT: In sediment-heavy sections of the **Yangtze River**, fish depend more on vibration than visibility. Suspended particles distort light but carry chemical and pressure signals efficiently. Murkiness becomes an advantage for species tuned to movement. Information rides the current.

MB FACT: In oxygen-poor lakes of **Finland**, **burbot** remain active beneath winter ice while most fish slow down. They detect seasonal chemical shifts that signal safe spawning windows. Cold water doesn't silence signals -- it sharpens them. Winter flips the rules.

MB FACT: In the **Mississippi River Delta**, **paddlefish** locate plankton using **electroreception** *(sensing weak electrical signals from living organisms)*. Their long snouts act like antennae, detecting prey without seeing it. Feeding becomes passive but precise. Freshwater hums electrically.

MB FACT: Isolated lakes in **Alaska** have trapped fish populations for **thousands of years**. Over time, local vibration patterns and chemical cues shape unique behaviors. Two fish of the same species may respond differently to identical signals. Geography rewrites instincts.

MB FACT: During spring floods in the **Mekong River,** migrating fish follow chemical trails rather than landmarks. These signals persist even when channels shift. Water remembers paths long after maps fail. Migration is guided by invisible lines.

MB FACT: In deep sections of **Lake Michigan, freshwater drum** generate low humming sounds by vibrating muscles against their swim bladder. Entire groups can produce overlapping noise fields during spawning season. Lakes are not silent -- they broadcast.

MB FACT: Mudminnows thrive where most fish fail. In swampy channels of **Atchafalaya Basin,** they sense oxygen loss early and surface to breathe air. Their name hides their toughness. Small bodies can carry extreme adaptations.

MB FACT: Fish don't just react to floods -- they remember them. After major surges in the **Ganges River**, lingering chemical and structural signals shape behavior for months. Routes avoided once may stay avoided forever. Rivers store memory.

MB FACT: Freshwater mussels quietly control water quality. In tributaries of the **Tennessee River**, a single mussel can filter **gallons per day**, altering chemical signals fish rely on. Silent animals shape the soundscape indirectly. Not all signals make noise.

MB FACT: Ancient body plans survived because they listen well. **Gar** and **sturgeon** respond effectively to pressure and vibration cues that newer designs struggle to interpret. In rivers flowing toward the Gulf of Mexico, old designs still read modern signals. Evolution rewards awareness.

MB FACT: Artificial lakes force fast behavioral rewrites. Fish must relearn how sound, flow, and vibration behave in man-made basins. In **Lake Mead**, behavior shifts appear long before bodies change. Freshwater evolution starts with listening.

MB FACT: Sea robins don't just swim -- they **walk across the seafloor**. Their wing-like fins hide finger-shaped legs that tap, probe, and "taste" the sand for buried prey. Each step sends sensory signals back to the brain, turning the ocean floor into a map you can feel. To a sea robin, the bottom isn't empty -- it's searchable.

On a Lighter Side

MYTHS - BUSTED

MYTH: Knocking sounds beneath frozen lakes were spirits trying to escape.
In regions of **Finland and Karelia**, people believed that rhythmic knocking heard through winter ice came from spirits trapped below the surface. The sounds were said to grow stronger before storms or thaws, as if something was pushing upward. Fishermen avoided drilling near the noise, fearing release. Ice that spoke was believed to be dangerous.

MYTH: Low humming in rivers warned of disaster ahead.
Boatmen along the **upper Rhine River** believed that a deep hum rising from the current was a warning from the river itself. The sound was said to appear shortly before accidents, capsizing boats or pulling swimmers under. Ignoring the hum was considered foolish. The river was thought to speak only when danger was close.

MYTH: Whistling near lakes summoned unseen beings below.
In **Scottish Highland lochs**, whistling was avoided because it was believed to call something up from beneath the water. When echoes or answering sounds followed, people fled the shoreline. The loch was said to hear and respond, but never kindly. Sound was treated as an invitation.

MYTH: Repeating splashes marked the movement of a hidden lake creature.
Early settlers around **Lake Champlain** described evenly spaced splashes that circled boats without breaking the surface. These sounds were believed to trace the path of a long, unseen

creature moving just below the water. When the splashing stopped, fear increased. Silence meant it was directly underneath.

MYTH: Rivers could mimic human voices to lure people in.

In Slavic folklore near the **Vistula River**, voices rising from the water at dusk were believed to imitate names, laughter, or familiar calls. Following the sound was said to lead people into deep currents. Those who returned claimed the voice changed once they approached. Water that spoke was never human.

MYTH: Singing lakes held the voices of the dead.

Among communities in **northern Siberia**, low tones rising from frozen lakes were believed to be the voices of those lost beneath the ice. The sounds were strongest at night or during sudden temperature shifts. Hunters avoided camping near these lakes. The water was thought to remember everyone it took.

MYTH: Repeating ripples meant the lake was aware of you.

In **Anishinaabe** stories from the Great Lakes region, evenly spaced ripples were believed to signal that the water had noticed a person. Each ripple marked time rather than movement. Leaving before the pattern ended was considered wise. Being counted by the lake was dangerous.

LEGENDS

REAL LEGENDS

The Shadow That Followed the Canoe
Lake of the Woods -- Canada / Minnesota

Paddlers crossing quiet bays reported a long shadow keeping pace beneath their canoe. It stayed just off the bow, never breaking the surface, matching speed without splashing. When they stopped, it stopped. When they turned, it slid away and reappeared on the other side.

Old fishing guides said the lake "likes to travel with you." Some blamed a single massive fish. Others said it was many fish moving together. Later studies showed large northern pike and sturgeon often cruise below boats, using the shadow as cover while investigating vibrations.

Those who experienced it remember the precision -- not fear, but awareness. Something was there, measuring, deciding. The lake didn't reveal what it was, only that it noticed them.

The River That Went Silent
Upper Mississippi River -- Midwestern United States

During late summer, river towns noticed something wrong. Nets came up empty. Surface ripples vanished. Even insects seemed fewer. Locals said the river had "gone quiet."

Fishermen warned newcomers not to waste bait. Old timers said it had happened before. Weeks later, fish returned suddenly, as if nothing had happened.

Biologists eventually connected the silence to low oxygen caused by warm temperatures and slow flow. Fish moved deeper or downstream temporarily. But for those who lived along it, the legend remained -- the river didn't die. It simply held its breath.

The Lake That Remembered Nets
Lake Geneva -- Central Europe

Fishermen complained that fish avoided certain areas no matter what bait they used. Nets were cut, lines snapped, hooks ignored. They said the lake had learned.

Older stories claimed the fish remembered traps from generations before. Scientists later discovered carp and other species can learn and pass avoidance behavior socially. Areas heavily netted in the past produced smarter fish.

The legend survived because it felt personal. The lake wasn't empty. It was selective. And it hadn't forgotten how it was hunted.

The Winter Fish That Woke the Lake
Great Slave Lake -- Northern Canada

In the dead of winter, when most lakes fall silent, fishermen reported sudden movement beneath the ice. Lines jerked unexpectedly. Sounds traveled far under frozen water. Something was active when nothing else should have been.

Local stories spoke of fish that belonged to winter, not summer. Scientists later confirmed burbot become more active in cold, dark conditions and spawn under ice.

Still, the timing unsettled people. When the lake should have slept, it didn't. It waited for cold -- then woke up.

The Shape That Surfaced at Dusk
Lake Champlain -- Northeastern United States

Boaters reported seeing long, rolling shapes just below the surface at sunset. The movement was slow and deliberate, disappearing when approached. Sightings clustered in low light, especially during calm evenings.

Some claimed a monster. Others said it was imagination. Research later pointed to schools of large fish moving together, surface distortion, and long-bodied species magnified by low-angle light.

The legend endures because of consistency. Same time. Same conditions. Same shape. The lake offers just enough clarity to keep the question alive.

MYTH: Loud voices over rivers angered water spirits.
In medieval **Central European** river lore, shouting or singing over water was believed to provoke unseen forces below. Sudden pulls, shifting currents, or overturned boats were blamed on disrespectful sound. Silence was seen as protection. The river listened before it acted.

DID YOU KNOW ?

Did You Know? Walleye and **burbot** don't just survive darkness -- they hunt with it. In northern lakes across **Minnesota, Wisconsin,** and **Ontario (Canada),** these fish become more active after sunset because their eyes are tuned to collect the faintest scraps of light and motion. As vision fades for others, **sound and vibration take over,** giving them a quiet edge. To them, night isn't a weakness. It's a hunting window.

Did You Know? Crayfish can sense danger without ever seeing it. In fast-moving rivers like the **Mississippi River (United States)** and the **Danube River (Central Europe),** a single tail flick upstream sends pressure waves racing through the water. Other crayfish react instantly -- freezing, darting for cover, or vanishing into rocks. For them, water isn't just flowing past. It's constantly sending warnings.

Did You Know? Freshwater fish often live closer to their physical limits than most ocean fish. A lake can lose oxygen in hours during a heat wave, lock it away under winter ice, or flip its chemistry overnight after an algae collapse. Fish that survive don't rely on strength -- they rely on timing, slowing their bodies, changing depth, or gulping air at just the right moment. From above, the water looks peaceful. Below, it's a nonstop survival puzzle.

Did You Know? Fish don't simply vanish -- they follow invisible borders. Where warm and cold water meet, oxygen and temperature line up just right, creating narrow **"sweet zones"** packed with life. One day fish stack tightly there; the next day they're gone without a trace. The fish didn't leave. **The**

boundary moved. Lakes rearrange life constantly without ever changing their surface.

Did You Know? In cloudy rivers and dark lakes, **smell and vibration** often matter more than sight. Scents don't just drift -- they form trails that bend, pool, and stretch with the current. Some fish can follow these chemical paths upstream with accuracy that feels intentional. Information moves through water even when nothing can be seen. Vision is optional. Signals are not.

Did You Know? Winter doesn't shut freshwater life down -- it rewrites the rules. Beneath ice, light disappears and oxygen can fall to dangerous levels, yet some fish become more active while others slow nearly to a standstill. Feeding roles flip, predators change schedules, and entire food webs rearrange. This is why one species bites aggressively in winter while another seems to vanish. Cold doesn't mean dead water. It means a different world.

Did You Know? The same fish species can behave like a completely different animal depending on where it lives. A **bass** in a clear, rocky lake feeds, moves, and reacts differently than one in a muddy river. These aren't genetic changes -- they're learned habits built over time. Fish remember what works where they live. Every lake and river trains its residents.

Did You Know? Many freshwater "monster" stories likely began with ordinary animals behaving under strange conditions. Murky water stretches shadows, surface wakes exaggerate length, and poor depth cues erase scale. A long fish swimming just below the surface can look massive, fast, and wrong. Add low light and a startled observer, and reality bends fast. Freshwater doesn't need new creatures to create legends -- it reshapes familiar ones.

STORY MOMENT

The Fish That Didn't Leave

The lake looked empty.

Not dead. Not abandoned. Just... vacant.
The kind of empty that messes with you. Three straight days of steady casting had produced nothing but silent lines, drifting weeds, and the faint slap of water against the hull. No strikes. No follows. No signs. Locals shrugged and said the fish had "moved on," driven deeper by the heat, pushed somewhere else entirely.

But when the old man at the launch ramp heard that, he shook his head slowly.

"They didn't leave," he said. "They changed."

That evening, the lake flattened as the sun sank low and turned the water copper and gray. The breeze died. The surface went glassy. No surface strikes. No chasing baitfish. Not even a nervous ripple. Just stillness stretched tight, like the lake was holding its breath.

Then -- right at the edge of the reeds -- the line twitched.

Once.
Twice.
Then slack.

Whatever touched it didn't run. It didn't panic or pull. It tested. Felt the tension. Measured resistance. Then let go. A contact without commitment. A decision made in the dark water below.

Later, the sonar told the real story. Fish weren't scattered. They were stacked -- tight and deliberate -- along a thin band of water just off the bottom. A narrow layer where oxygen held longer and temperatures stayed steady. They weren't roaming the lake anymore. They weren't hunting wide.

They were waiting.

A largemouth finally took the bait near dusk. Not a chase. Not a splash. A sudden, precise hit felt more than seen. It struck by vibration, by pressure, by instinct refined through heat and silence. When it surfaced, it wasn't big -- just controlled. Purposeful. This fish hadn't rushed. It had calculated.

As darkness settled, the lake began to move again.

Drum hummed faintly below the surface, a low vibration you felt more than heard. Carp rolled near submerged timber, breaking the skin of the water and vanishing again. Somewhere deeper, something large shifted once -- slowly, deliberately -- just enough to bend the water and disappear.

By morning, the story had already changed. People said the fish had "come back." Said the bite was on again. Said the lake had woken up.

The old man only smiled.

They were never gone.

The lake hadn't emptied. It had rearranged itself. Layers had shifted. Rules had changed. And the fish -- patient, adaptable, aware -- had simply waited for the water to make sense again.

FUN QUIZ

1. **What is the main survival advantage described across many freshwater species in this chapter?**

   A. Stronger muscles for fast swimming

   B. Perfect adaptation to one stable environment

   C. The ability to detect rapid, invisible environmental changes

   D. Larger brains for learning new routes

2. **Why can largemouth bass hunt successfully in muddy Mississippi River backwaters?**

   A. They use echolocation like dolphins

   B. They rely on smell alone

   C. They detect pressure waves using their lateral line

   D. They wait for clearer water before feeding

3. **How do catfish in the Ohio River "see" their surroundings in dark water?**

   A. By detecting temperature gradients

   B. By tasting chemical signals with their skin and whiskers

   C. By using reflected moonlight

   D. By following sound echoes

4. **What role does mud play in how crayfish detect danger in some streams?**

A. It blocks predators from approaching

B. It amplifies sound waves in open water

C. It transmits low-frequency vibrations through sediment

D. It stores chemical signals permanently

5. Why do cave-dwelling fish often lose their eyesight over generations?

A. Eyes are damaged by mineral-rich water

B. Darkness makes vision unnecessary, so energy shifts to other senses

C. Predators target fish with eyes

D. Light sensors interfere with chemical detection

6. What allows fish to escape oxygen crashes in lakes like Lake Winnipeg or Lake Erie?

A. Faster swimming speeds

B. Larger gills

C. Early detection of chemical and behavioral warning signals

D. Migration to deeper water only

7. How do species like northern pike succeed as predators without long chases?

A. By overwhelming prey with group attacks

B. By tracking prey visually over long distances

C. By timing ambush strikes using precise vibration patterns

D. By exhausting prey slowly

FUN QUIZ ANSWERS

1. C – The chapter emphasizes survival through sensing rapid, invisible environmental change rather than strength or perfect adaptation.

2. C – Largemouth bass detect pressure waves using the lateral line system, allowing them to hunt without relying on sight.

3. B – Catfish rely on chemical sensing through their skin and whiskers, effectively "tasting" the water around them.

4. C – Vibrations travel through sediment, letting crayfish sense danger through the riverbed before predators appear.

5. B – In permanent darkness, energy shifts from eyesight to pressure and chemical sensing, making vision unnecessary.

6. C – Fish survive oxygen crashes by detecting early chemical and behavioral warning signals and escaping in time.

7. C – Northern pike time ambush strikes by reading precise vibration patterns instead of chasing prey.

CHAPTER 7

7-MONSTERS & LEGENDS...Maybe True?
Where Strange and Real Become One

FUN & WEIRD FACTS

MB Fact = Mind-Blown Fact
Real discoveries. Real mysteries, Real Facts. Fun to make you think.

MB FACT: Off the coast of **Newfoundland, Canada, fin whales** -- the second-largest animals on Earth -- can reach **80–85 feet** long. When one surfaces in rough seas, only sections of its back appear between waves. Witnesses often reported a creature "longer than the ship," because the visible pieces surfaced in sequence. The animal was real. The length felt impossible.

MB FACT: In **Lake Superior (United States–Canada)**, waves can exceed **25 feet** during fall storms, taller than a two-story house. When large fish or submerged objects move beneath these waves, the crests rise and collapse unevenly. Early sailors described "backs" rolling under the surface. The lake hid scale inside scale.

MB FACT: Along the **Norwegian Sea**, **sperm whales** can dive over **7,000 feet**, deeper than most mountains are tall. When they surface suddenly after long dives, the force displaces massive volumes of water. From a distance, the surge looks like something huge pushing upward from below. Depth turns resurfacing into spectacle.

MB FACT: During violent storms in the **North Atlantic shipping lanes between Newfoundland, Canada and Ireland**, sailors repeatedly overestimated the size of animals breaking the surface. Rolling decks, crashing waves, darkness, and fear destroy depth perception. A whale or large fish surfacing briefly between waves can appear **many times larger** than reality. Old ship logs from the 1700s and 1800s often list wildly different lengths for the same encounter. The sea doesn't shrink creatures -- it stretches perception.

MB FACT: For centuries, the **giant squid** was dismissed as legend despite reports from sailors in the **Norwegian Sea and North Atlantic**. No intact specimen was confirmed by science until the **mid-1800s**, when bodies finally washed ashore. Even today, live sightings in the deep remain rare and fleeting. Something long believed impossible turned out to be real -- just hidden where humans rarely look. Some legends survived simply because they were right too early.

MB FACT: Long-bodied animals create powerful illusions near the surface in places like the **Sea of Japan and the Bay of Biscay**. Only sections of the body appear between waves while the rest remains hidden below. The brain fills in the gaps, often exaggerating length and continuity. A surfacing pattern becomes a "serpent" in memory. Water turns fragments into monsters.

MB FACT: Lake monster sightings often cluster at specific times and places, such as **Loch Ness in northern Scotland near Inverness**, especially during calm evenings with low-angle light. These conditions flatten distance cues and distort speed. When familiar shorelines remove reference points, movement appears larger and closer than it is. Repeated sightings don't require a single creature. They require the same conditions repeating.

MB FACT: Early sailors encountering unfamiliar animals in the **South Atlantic and Indian Oceans** had no scientific framework to interpret what they saw. When strange shapes surfaced, artists emphasized features that stood out -- **eyes, coils, arms** -- while ignoring scale and context. These exaggerated drawings spread faster than corrections ever could. Over time, emphasis became anatomy. The legend took shape before the animal did.

MB FACT: Some large marine animals behave in unsettling ways when sick, dying, or disoriented, especially along **whaling routes near Greenland and Iceland**. They may surface repeatedly, thrash near ships, or drift unnaturally close to humans. These rare moments are far more likely to be witnessed than normal behavior. Legends tend to form from abnormal encounters, not everyday life. Monsters often begin at the edge of death.

MB FACT: In the **Amazon River near Santarém, Brazil**, migrating fish schools can stretch for **hundreds of meters**. As they move together just under the surface, the river bulges and smooths in wide bands. Early explorers described the water rising "like a single moving body." Thousands of animals collapsed into one shape.

MB FACT: In the **Bay of Biscay off France and Spain**, underwater landslides can shift millions of tons of sediment at once. These events create surface disturbances that travel fast and then vanish completely. To witnesses, the water seemed to lift and flee. No body remained. Just the feeling of something enormous moving away.

MB FACT: Oarfish, recorded off **Japan and California**, can exceed **35 feet** in length and often swim vertically. When sick or disoriented, they surface in pieces -- head first, then body, then tail. Observers rarely see the whole animal at once. The result looks like a rising serpent with no end.

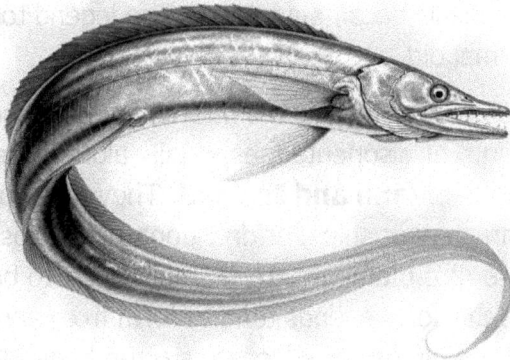

MB FACT: The human brain is wired to find patterns under threat, especially in low visibility. In foggy waters like the **Grand Banks off Newfoundland**, scattered movement quickly connects into familiar shapes. Water hides edges and scale, letting imagination finish the outline. Once a pattern forms, memory reinforces it. The monster appears fully formed after the moment has passed.

MB FACT: Reports of mermaids peaked after long voyages across the **Mediterranean Sea and Caribbean trade routes**. Exhaustion, dehydration, and isolation distort perception, especially when land has been absent for weeks. **Manatees and seals** surfacing briefly at a distance can appear eerily human under those conditions. The sight was real. The interpretation was fragile.

MB FACT: Some legends survived simply because no one ever returned to challenge them. Remote lakes and ocean regions like

Lake Baikal in Siberia, Russia, were visited once and never again by the same witnesses. Without repeated observation, stories hardened instead of being tested. Silence protected belief. Distance preserved mystery.

MB FACT: Massive wakes near shorelines in places like the **Gulf of St. Lawrence, Canada**, can be created by large schools of fish moving together just beneath the surface. From land, the motion looks like a single, powerful body traveling with intent. When the wake vanishes suddenly, it feels deliberate. Group behavior collapses into one creature in the human mind.

MB FACT: In **Glacier Bay, Alaska (United States)**, entire **stretches of water have been reported to "boil"** without warning. The effect comes from large numbers of seals, fish, or whales surfacing and diving together beneath cold, glassy water. To someone watching from shore, the surface heaves like something enormous is rolling just below. No body appears. Just motion. That kind of movement doesn't feel accidental.

MB FACT: Along the **Bay of Fundy in eastern Canada**, extreme tides can **create long, fast-moving surface bulges** that race parallel to shore. These bulges hold their shape for minutes before collapsing. Early witnesses described them as traveling backs or spines. When the water suddenly flattened, it felt like whatever was there chose to disappear.

MB FACT: In deep lakes like **Lake Champlain on the border of the United States and Canada**, long fish surfacing briefly can appear to rise vertically rather than horizontally. Without waves or horizon cues, the brain misreads orientation. A single roll can look like a neck lifting out of the water. The movement is real. The scale is not.

MB FACT: Sailors in the **South China Sea** frequently reported hearing splashes that sounded too heavy for any known fish. Large rays and sharks can slap the surface with fins or tails when feeding, producing sharp, rhythmic impacts. When the body stays hidden and only the sound repeats, presence replaces proof. Something is clearly there. Just not fully seen.

MB FACT: In rivers like the **Amazon near Manaus, Brazil**, massive schools of fish can move upstream just beneath the surface, pushing water ahead of them. The surface rises smoothly, then drops all at once when the school dives. From a canoe, the water itself seems alive. Early travelers described these moments as the river "breathing."

MB FACT: In calm stretches of the **Mississippi River (United States)**, large sturgeon weighing over **200 pounds** can roll near the surface while feeding. Their armored backs briefly break the water before slipping under again. Seen at dusk, the movement looks deliberate and massive. Armor reads as intention.

MB FACT: Along **Antarctic waters near the Drake Passage**, ice breaking underwater can release shock-like pulses felt through ship hulls. These impacts arrive without warning and leave no visible source. Early crews described being struck from below. The sound came first. Fear filled in the shape.

MB FACT: In **Loch Morar, Scotland**, one of Europe's deepest freshwater lakes at over **1,000 feet**, objects surfacing from depth take longer to rise and longer to sink. A single movement can feel slow, heavy, and controlled. Witnesses reported shapes that "meant to surface." Depth changes timing -- and timing changes belief.

MB FACT: In the **Gulf of California (Sea of Cortez), Mexico**, massive schools of mobula rays can leap from the water by the hundreds. When they land nearly together, the impacts echo and churn the surface violently. From shore, the splashes blur into one overwhelming presence. Many bodies become one event.

MB FACT: Some whale species roll sideways when surfacing, exposing a long curve of body before slipping back under. In fog-heavy regions like the **North Pacific near the Aleutian Islands**, only sections are visible at a time. The surfacing points don't line up. The eye connects them anyway. One animal becomes many coils.

MB FACT: In calm bays such as **San Francisco Bay (United States)**, reflected city lights at night can break apart on ripples and wakes. When something large moves just below the surface, the reflections stretch and bend into glowing shapes. Witnesses described moving eyes or lit backs sliding through the water. Light turns motion into something watchful.

MB FACT: Along the **St. Lawrence River (Canada)**, large fish and marine mammals often travel just beneath moving boats, using shadow and vibration for cover. When the boat turns, the shadow turns. When it slows, the follower slows. To someone leaning over the rail, it feels deliberate. As if something is choosing to stay close.

MB FACT: Early sonar operators in **World War II Atlantic patrol zones** reported contacts that rose, sank, and curved away from ships. Some were later linked to schools of fish or thermal layers, but not all could be recreated. On primitive displays, moving echoes looked alive. A blip that reacts feels different than one that doesn't.

MB FACT: Many legendary creatures are described as appearing once, then never again. In remote waters like **Lake Baikal in**

Siberia, Russia, witnesses often had no second chance to observe what they saw. Without follow-up encounters, the moment stands alone. A single unexplained sighting can carry more weight than a hundred ordinary ones.

MB FACT: In the **North Atlantic south of Greenland**, rogue waves over **60 feet tall** have been recorded -- taller than a **six-story building** rising out of nowhere. If a whale or shark surfaces near one, the wave can briefly lift part of its body upright. To sailors, it looked like a creature standing out of the sea. The animal wasn't changing shape. The ocean was.

MB FACT: Sperm whales produce clicks measured at over **230 decibels**, louder than a **rocket launch at close range**. In deep waters near the **Azores in the Atlantic**, these sounds can travel miles and vibrate metal hulls. Early crews reported hammering from below, even when the sea was calm. Hearing something that powerful without seeing it feels like being targeted.

MB FACT: Lake Tanganyika in East Africa is over **4,700 feet deep**, deeper than the tallest skyscrapers are tall. When gases shift in its deep layers, entire sections of water can move strangely and force fish to the surface all at once. To shoreline villages, the lake seemed to rise and exhale. Water that deep isn't supposed to move like that.

MB FACT: Along the **Pacific coast of Chile**, giant kelp forests can grow over **100 feet long**, about the length of a **blue whale**. During storms, torn kelp rises and sinks in slow, sweeping motions beneath the surface. From a small boat, the movement looks like coils sliding just below the water. Plants borrowed the shape of monsters.

MB FACT: In the **Ganges River delta of India and Bangladesh**, massive schools of fish can erupt simultaneously when predators strike below. Tens of thousands of bodies break the surface at once, then vanish in seconds. Witnesses described the river attacking itself. When motion stops that fast, it feels intentional.

MB FACT: Whale sharks, the largest fish alive, can exceed **40 feet long** and weigh more than **20 tons** -- about the size of a **school bus**. Near the **Yucatán Peninsula**, they often feed just under the surface, revealing only sections of their backs. One pass can look like several creatures surfacing in sequence. Size becomes confusing when you never see the whole thing.

MB FACT: Lake Baikal in Siberia, Russia plunges more than **5,300 feet**, deeper than **one mile straight down**. Objects moving there can drop out of sight almost instantly. Sonar has recorded fast-moving shapes that vanish vertically instead of swimming away. To witnesses, it felt like something choosing depth as an escape.

MB FACT: In the **Bering Sea**, massive ice slabs can break underwater and roll upward without immediately breaking the surface. The water bulges slowly, then collapses with force. Crews described being followed by something rising beneath them. Ice moving silently feels deliberate.

MB FACT: Along the **coast of South Africa, groups of dolphins** can herd fish into tight spinning formations near the surface. The rotating water can span **hundreds of feet**, wide enough to surround boats. From above, it looks like one enormous body circling. Coordination creates scale.

On a Lighter Side

*"I've been put on hold.
Something to do with all that plastic
you folks have been throwing overboard"*

"I can't believe the treasure and plants are fake!
Guys... I think we're in an aquarium!"

MYTHS - BUSTED

MYTH: In **the Bermuda Triangle between Florida, Bermuda, and Puerto Rico,** sailors believed time itself could slip. Crews reported watches stopping, restarting, or showing different times on the same ship. The belief wasn't that something attacked vessels, but that entering the region meant leaving normal time behind -- and ships that stayed too long simply never reached the same moment again.

MYTH: In **the Dragon Triangle south of Japan near Iwo Jima,** fishermen believed the sea could erase direction. When compasses failed and stars no longer lined up, it was said the water had no "up" or "forward." Ships were thought to drift endlessly without sinking, lost not to depth but to orientation.

MYTH: On **Lake Michigan near the Wisconsin coast,** mariners believed the lake could hide wrecks by swallowing evidence, not ships. Vessels were said to vanish without debris, oil, or bodies because the lake "closed" behind them. The fear wasn't monsters -- it was disappearance without trace.

MYTH: On quiet nights along **the Hudson River north of West Point,** boaters believed the water could suddenly reverse direction without warning. Oars pulled one way while the current dragged another. The belief was that the river remembered old channels beneath the surface and occasionally followed them again, taking anything floating along with it.

MYTH: In **the Black Sea along the Turkish coast,** sailors believed certain depths were cursed because nothing living could survive there, yet sounds still came from below. Anchors lowered too far were thought to disturb places meant to stay untouched. Silence afterward was taken as a warning, not relief.

MYTH: Along **the coast of Namibia in the Skeleton Coast**, fogbanks were believed to move against the wind on purpose. Ships entering the fog were said to lose the horizon entirely, unable to tell sea from sky. Crews feared becoming trapped in a white world with no direction, drifting until supplies ran out.

MYTH: On **Lake Baikal in Siberia**, locals believed the lake rejected intruders who crossed certain invisible lines. Boats that stalled or lost power in calm water were thought to have crossed a boundary where the lake decided who could pass. Survivors spoke of sudden panic without knowing why.

MYTH: In **the South China Sea near the Paracel Islands**, sailors believed the water could fake land. Dark shapes appeared on the horizon, growing clearer as ships approached, only to dissolve completely. Crews feared steering toward safety that wasn't real, running out of fuel chasing solid ground that never existed.

MYTH: Along **the River Thames near Richmond**, rowers believed the water could suddenly go silent, muting splashes and oar sounds even in daylight. Locals warned that when the river stopped echoing, it meant something was wrong with the current itself, and staying on the water too long would leave boats drifting without realizing it.

MYTH: Fishermen on **Lake Maracaibo in western Venezuela** believed lightning didn't always come from the sky. Sudden flashes over calm water were said to rise from the lake itself, marking places boats shouldn't linger. The warning wasn't about storms -- it was about the lake choosing where energy escaped.

MYTH: In **the Danube near the Iron Gates gorge between Serbia and Romania**, crews believed the river could steal momentum. Boats moving downstream would suddenly slow or stop as if pushing against invisible resistance. Locals said the river tightened there, and fighting it only made things worse.

MYTH: Near **the Baltic Sea around the Åland Islands**, fishermen believed sound could mislead. Bells, voices, or knocks were said to carry from the wrong direction, drawing boats toward reefs. Following sound instead of sight was considered a deadly mistake passed down through generations.

MYTH: On **Lake Titicaca between Peru and Bolivia**, it was believed the lake punished disrespect. Sudden storms were blamed on laughter, shouting, or careless behavior on the water. The myth warned that some places weren't dangerous because of animals, but because they demanded silence.

MYTH: Off **the coast of Newfoundland near the Grand Banks**, sailors believed the sea could mimic distress. Phantom lights, false flares, and imagined calls were said to appear during heavy weather, luring ships off safe routes. Crews argued whether the danger was deception -- or the human mind under stress.

MYTH: When the **USS Eldridge** docked in **Philadelphia Harbor in 1943**, sailors believed the ship briefly slipped out of normal reality during a secret naval experiment. Crew members later whispered that the ship vanished in green haze, reappeared miles away, and returned with men missing, sick, or fused into the deck. The belief was that the ocean could be bent by human technology -- and that water remembers when it happens.

MYTH: In **the Black Sea off the coast of Turkey**, sailors believed lowering anchors past a certain depth disturbed places meant to stay sealed. When bubbles or surface boils appeared afterward, crews cut lines and fled. The belief held that some depths were not part of the living world anymore.

MYTH: Submarines operating near **the Baltic Sea during World War II** reported phantom sonar contacts that followed without closing distance. Operators believed these were not vessels but echoes that learned. Once detected, they stayed until the ship left the area -- or shut down entirely.

MYTH: In **Lake Superior near Whitefish Point**, crews believed the lake chose when to claim ships. Sudden storms were said to rise without warning, swallowing large freighters whole. Survivors warned that Superior didn't sink ships slowly -- it erased them.

MYTH: Pilots flying low over **the Bass Strait between Australia and Tasmania** believed the water could pull aircraft downward without visible weather. Engines lost power, gauges spun, and planes vanished mid-route. Locals said the strait didn't just swallow ships -- it reached upward.

MYTH: Fishermen on **Lake Lanier in northern Georgia** warned that submerged roads and structures caused boats to stall even in calm water. Engines were said to fail directly over certain points, as if the lake itself resisted being crossed. Locals believed the water wasn't dangerous everywhere -- only where the land still existed below.

MYTH: Along **the Mississippi River near Baton Rouge**, barge crews believed the river could fake depth. Poles and lines gave false readings, making safe water suddenly vanish beneath a hull. The warning wasn't about sinking -- it was about trusting measurements that lied.

MYTH: In **Lake Okeechobee in southern Florida**, sudden stillness was feared more than storms. When wind died completely and birds went silent, locals believed the lake was about to release something violent. Boats rushed for shore, convinced the calm meant pressure building below.

MYTH: Kayakers on **the Snake River in Idaho** believed certain bends caused time confusion. Trips took longer or shorter than expected, with daylight fading too fast or not fast enough. The belief was that the canyon trapped sound and light, making judgment unreliable.

MYTH: Near **the Chesapeake Bay's Tangier Sound**, watermen warned never to follow voices across open water at dusk. Calls were said to carry unnaturally far and from the wrong direction. The belief wasn't that someone was calling -- it was that the bay echoed intentions, not sound.

MYTH: Along **the coast near Whitby in northern England**, sailors believed fog could lower itself deliberately, cutting ships off from landmarks they could see moments earlier. When cliffs vanished without wind, crews turned back immediately, fearing the fog wasn't weather but a boundary.

MYTH: On **Lake Geneva near Montreux**, boaters warned against crossing certain open-water lines at dusk. Reflections of mountains were said to bend depth and distance so badly that people misjudged direction entirely. The belief was simple: follow reflections and you'd end up nowhere.

MYTH: In **the canals of Amsterdam during winter**, locals believed ice could move without cracking. Boats trapped overnight were found shifted from where they had frozen in, as if the water had flowed beneath solid surfaces. People said the canals never truly stopped moving.

MYTH: On **Lake Mead near the old submerged town sites**, divers believed panic came without cause. Perfectly clear water suddenly felt suffocating, forcing ascents without explanation. Locals said the lake rejected people who stayed too long where buildings still stood.

MYTH: Along **the Columbia River Gorge**, boaters believed wind could move underwater first. Sudden sideways pulls happened before gusts reached the surface. Those who ignored it were said to lose control before realizing conditions had changed.

MYTH: In **the Finger Lakes region of New York**, residents believed long lakes created false distance. Objects on the water appeared close but took hours to reach. The warning was simple:

if something looks reachable, it probably isn't -- and trying might leave you stranded.

MYTH: Near **Lake Pontchartrain outside New Orleans**, locals believed the lake could hide storms. Dark water stayed calm while violent weather formed just beyond sight. Boats that waited for visible signs were said to be caught with no time to escape.

MYTH: Near **the Yellow River outside Lanzhou in China**, ferrymen believed the water could change weight. Boats rode lower without taking on water, forcing crews to unload cargo suddenly. The river was said to thicken without warning, making it dangerous to cross even in calm weather.

MYTH: On **Lake Balaton in western Hungary**, sudden wind walls were feared more than storms. Locals said the lake could switch directions instantly, pushing small boats toward shore faster than oars could counter. When the water darkened without clouds, people got off immediately.

MYTH: Along **the Ganges near Varanasi**, boatmen believed sound could travel ahead of action. Bells or chanting were sometimes heard before boats appeared. Following the sound was considered unlucky, as if the river was warning you too early on purpose.

MYTH: In **the fjords near Bergen, Norway**, residents believed still water meant pressure building, not safety. Perfect reflections were taken as signs to stay ashore, because sudden downdrafts and water movement could arrive without surface warning. Calm wasn't trusted -- it was watched.

REAL LEGENDS

LEGENDS

The Kraken's Wake
North Atlantic Ocean -- off the coasts of Norway and Iceland

Sailors told of a calm sea suddenly rising beneath their ships, as if the water itself were lifting. Nets vanished. Oars snapped. Dark shapes moved just below the surface without fully appearing. What terrified crews most was not an attack, but the *stillness* that followed -- as if the sea had decided whether to let them go.

Modern science links these accounts to giant squid and massive schools of deep-water animals rising unexpectedly. Yet even today, live observations of giant squid in their natural environment are rare. The encounters were brief, chaotic, and impossible to repeat. Whether the sailors saw one creature or many, something large was undeniably present. The sea gave no second look.

The Lake That Follows You
Lake Champlain -- Northeastern United States

Boaters reported long shadows pacing their vessels just beneath the surface. The movement stayed parallel, never surfacing fully, never falling behind. When the boat slowed, the shape slowed too. When the engine stopped, the water stilled -- except for a slow roll beneath the hull.

Researchers later noted that large fish often follow boats, attracted by vibration and shadow. Yet witnesses insist this was different. The movement felt deliberate, controlled, aware. The lake didn't just contain something -- it seemed to watch. Even skeptics admit the consistency of these reports is unsettling.

The Serpent of the Silent River
Upper Mississippi River -- United States

River pilots told of something long and dark surfacing in smooth arcs during low water. No splash. No thrashing. Just a slow rise, a bend, and disappearance. The creature was never seen head to tail, only in sections.

Biologists point to large sturgeon and coordinated fish movement as likely explanations. But the sightings clustered in specific bends of the river, often at dusk, often in silence. Those who saw it said the river felt "awake" afterward. The serpent was gone, but the sense of being observed lingered.

The Mermaids of the Warm Current
Caribbean Sea -- Near historical trade routes

Sailors described human-like figures resting on rocks or floating near the surface, watching ships pass. The figures vanished when approached, leaving only ripples. These sightings occurred most often after weeks at sea, during extreme heat and exhaustion.

Manatees and seals can briefly resemble human forms at a distance, especially in shimmering water. Yet some reports describe movement too upright, too deliberate. The question isn't whether sailors were wrong -- it's whether perception alone explains every account. Fatigue explains much, but not everything remembered so clearly.

On a Lighter Side

"Isn't this the life? No traffic, no noise...
just you, me and the calm ocean."

DID YOU KNOW ?

Did You Know?
In deep, cold lakes like Lake Superior, **lake trout** don't roam randomly. They track invisible temperature layers so precisely that a change of just a few feet can mean the difference between feeding and suffocating. During late summer, entire populations compress into thin underwater bands no thicker than a house ceiling. The lake may look vast and open, but livable space shrinks into a narrow ribbon. Survival becomes a matter of finding the right layer and never leaving it.

Did You Know?
In fast rivers, **freshwater mussels** can sense approaching fish before contact ever happens. They detect pressure changes and water movement through their soft tissue, snapping shut milliseconds before danger arrives. Some species even release larvae only when specific fish vibrations pass by, using the predator's presence as a delivery system. To the eye, mussels seem passive and still. In reality, they are constantly listening to the river.

Did You Know?
Some river fish survive drought by slowing life almost to a stop. **African lungfish** burrow into drying mud, secrete a cocoon of hardened mucus, and lower their metabolism so drastically they can remain alive for months without water. Their heart rate drops, movement ceases, and time itself seems to pause. When rains return, they rehydrate and swim away as if nothing happened. To an observer, it looks like resurrection rather than survival.

Did You Know?

In murky floodplain lakes of the Amazon, **electric knifefish** navigate using self-generated electric fields instead of sight. Every movement bends the field around their bodies, creating a living map of obstacles, prey, and rivals. When waters rise and forests flood, visibility drops to nearly zero, but their world becomes sharper, not darker. What looks like blindness is actually a different kind of vision.

Did You Know?

In northern lakes, **burrowing mayfly larvae** time their entire lives around oxygen crashes. They hatch, grow, and emerge as adults during brief windows when oxygen levels spike, sometimes lasting only days. Miss that window, and the next generation fails. Calm water hides this brutal schedule beneath the surface. What appears peaceful is governed by invisible deadlines.

Did You Know?

In shallow rivers, **freshwater stingrays** detect prey by sensing tiny electrical signals from muscle contractions. Even buried animals give themselves away with a heartbeat or twitch. The ray glides over sand that looks empty, then strikes with perfect accuracy. Sight plays almost no role in the hunt. The riverbed is alive with signals humans can't feel.

Did You Know?
Some lake fish experience sound as pressure, not noise.
Common carp can detect low-frequency vibrations from boats, storms, and distant shoreline movement through their swim bladder and inner ear connection. Long before ripples reach them, they respond by diving or freezing in place. To humans, the lake seems quiet. Underwater, it is full of warnings.

Did You Know?
In cold mountain streams, **stonefly nymphs** survive winter locked in ice-fringed water where temperatures hover just above freezing. Their bodies produce natural antifreeze compounds that prevent cells from rupturing. While the surface appears lifeless, activity continues below, slow but deliberate. What looks like a frozen pause is actually endurance in motion.

Did You Know?
In dark, tannin-stained lakes, **juvenile northern pike** rely on contrast rather than shape to hunt. They key in on slight changes in light caused by moving bodies, not outlines or color. Still prey blends into the background, but motion creates a flash of difference the pike cannot ignore. Camouflage fails the moment something moves. Stillness becomes a survival skill.

Did You Know?
Some freshwater predators don't chase at all. **Alligator gar** often float motionless near the surface, mimicking drifting logs for hours at a time. Their stillness lowers oxygen demand and hides them in plain sight. When prey passes underneath, the strike is sudden and explosive. What looks dead and harmless is actually waiting.

STORY MOMENT

STORY MOMENT

The Shape That Didn't Break the Surface

The lake was too calm. No wind. No chop. Just dark water reflecting the sky like glass. He'd fished here for years, but tonight felt wrong -- like the lake was holding its breath.

The first sign wasn't something he saw. It was something he felt. A low, hollow pressure pushed against the hull, heavy and slow. The boat rocked once, then steadied. He stood and scanned the surface. Nothing. No splash. No ripples.

Then a shadow slid beneath the boat.

It wasn't fast. It wasn't sharp. It was broad and steady, darker than the water around it. The fish finder flashed a thick mark, then went blank again. Whatever it was didn't chase or scatter. It passed, unhurried, as if it knew the boat was there.

He cut the motor. Silence stretched.

The shadow turned once, tracing the boat's outline, close enough to shift the water -- then faded into deeper darkness.

No one believed him. He didn't argue.

What stayed with him wasn't fear.
It was knowing something noticed him -- and chose to move on.

FUN QUIZ

---

1. **Which situation most often caused sailors to believe sea creatures were far larger than they really were?**

   A. Seeing animals during storms with waves and poor visibility

   B. Measuring animals while docked in calm harbors

   C. Watching animals from above in clear daylight

   D. Comparing sightings with scientific illustrations

2. **Why did long-bodied animals near the surface often appear serpent-like to witnesses?**

   A. Their scales reflected light in unusual colors

   B. Only parts of their bodies surfaced while waves hid the rest

   C. They swam in perfect circles near boats

   D. Their shadows moved faster than their bodies

3. **What made giant squid seem mythical for so long?**

   A. Sailors exaggerated their size on purpose

   B. They lived only in freshwater lakes

   C. No complete specimen was documented until the 1800s

   D. They were confused with whales

4. **Why do repeated monster sightings in the same location not always mean the same creature was seen?**

    A. Witnesses often copied earlier stories

    B. Similar lighting and water conditions can repeat the same illusion

    C. Creatures migrate in predictable patterns

    D. Legends required multiple sightings to survive

5. **Why did early sailors focus so heavily on eyes in monster descriptions?**

    A. Most sea animals have unusually large eyes

    B. Humans are biologically tuned to notice eye-like shapes first

    C. Reflections underwater always form circles

    D. Artists exaggerated eyes to make stories scarier

6. **How can schools of fish create the illusion of a single massive creature?**

    A. By swimming in perfectly straight lines

    B. By surfacing one at a time

    C. By moving together and shaping the water's surface

    D. By producing loud sounds simultaneously

FUN QUIZ ANSWERS

1. A
2. B
3. C
4. B
5. B
6. C

CHAPTER 8

8-MAPPING, EXPLORING & LIVING BELOW

Charting the Unknown. Going Where Nobody Has Gone

FUN & WEIRD FACTS

MB Fact = Mind-Blown Fact
Real discoveries. Real mysteries, Real Facts. Fun to make you think.

MB FACT: Sonar screens once showed **moving shapes that acted alive**. In the **1950s**, U.S. and Soviet naval crews tracking ships in the **North Atlantic** saw contacts that **matched their speed and gently changed course**, staying nearby for hours. The motion felt curious, not hostile. Years later, scientists confirmed many of these tracks were **whales swimming alongside ships**, using the noise as something to investigate. Early sonar was powerful -- but it didn't know the difference between steel and life.

MB FACT: Calm water suddenly formed **long ripples sliding across the surface**, with no splash and no visible source. This happened in **Scottish lochs such as Loch Fyne**, where residents noticed strange movement at night. Years later, records showed the **Royal Navy had been testing submarines** deep below. Steep underwater walls carried motion far from its source. From shore, it looked like something big was passing just under the skin of the water.

MB FACT: A shipwreck the size of a city block was found by following a trail of "snow." In **1985**, the search team led by **Robert Ballard** located the **Titanic** in the **North Atlantic** by spotting a long debris field first -- plates, coal, and twisted metal scattered across the seafloor. Once you find the "breadcrumb trail," the ship can't be far. The wreck sits about **2.4 miles down**, deeper than most mountains are tall, where sunlight never reaches. Deep exploration often starts with the scraps, not the headline.

MB FACT: Some "treasure hunts" are basically underwater math. Off the **Florida Keys**, the Spanish galleon **Nuestra Señora de Atocha** sank in **1622**, and searchers later used careful grid mapping to track where storms dragged its cargo. When **Mel Fisher's team** finally struck the main find in the **1980s**, it wasn't one chest -- it was a spread-out field of silver bars and artifacts. Hurricanes had "reorganized" the wreck for centuries like a giant invisible bulldozer. In the ocean, treasure doesn't sit still -- maps have to chase it.

MB FACT: Whole cities can vanish under the sea... and then show up again on sonar like ghost outlines. In **Aboukir Bay near Alexandria, Egypt**, underwater archaeologists mapped ruins of the ancient port city **Thonis-Heracleion**, including statues, temples, and stone blocks scattered across the seabed. From above, it's just water. Down below, it's streets and monuments lying sideways, half-buried like a toppled museum. Mapping turns "legendary lost city" into something you can measure.

MB FACT: Ancient shipwrecks can look like a spilled backpack -- then turn out to be a time capsule. Off **Turkey's southern coast near Kaş**, divers on the **Uluburun shipwreck** found thousands of objects packed together: copper ingots shaped like giant flat "pancakes," glass, ivory, and jewelry. It sank over **3,000 years ago**, and the cargo settled into the seafloor like a frozen shopping list from the Bronze Age. Mapping the pile mattered because one wrong fin-kick could scatter history. Underwater archaeology is treasure hunting -- but with rulers and patience.

MB FACT: The seafloor can erase footprints in a single day. In the **South China Sea**, researchers have documented **underwater sediment flows** -- muddy rivers that run downhill underwater -- capable of smoothing tracks and rearranging the bottom surprisingly fast. To a robot or diver returning later, the route can look "wrong," like the map got edited overnight. That's why deep expeditions log everything constantly. The ocean doesn't just hide things -- it can actively rewrite the scene.

MB FACT: Explorers don't "drive" deep-sea robots like cars -- they fly them like kites on a long leash. During wreck surveys in places like the **North Atlantic** and **Mediterranean**, ROVs (remotely operated vehicles) are tethered to ships by thick cables that carry power and video. Currents can push the robot sideways while the ship drifts above, so pilots make tiny corrections for hours. On screen, it looks like a slow glide over alien terrain. Deep mapping is part video game, part chess match with water.

MB FACT: Underwater navigation can work like throwing invisible "ping" breadcrumbs across the seafloor. In deep exploration zones like **Monterey Canyon off California**, teams place acoustic beacons that send timed pings so robots can calculate position -- basically **underwater GPS**, since satellite signals don't reach through water. A robot listens, triangulates, and updates its map as it moves. Without those pings, even a perfect camera can't tell you exactly where "here" is. Mapping the deep starts with finding your own dot on the map.

MB FACT: Some wrecks are so intact they look staged -- because the ocean can be a giant refrigerator. In the **cold waters of the Baltic Sea**, shipwrecks can survive with wooden details still visible because low temperatures and unusual conditions slow decay. Divers have described seeing decks, rails, and even cargo outlines that look shockingly "fresh" for something centuries old. The eerie part is how normal it can look -- like the ship is just waiting. In some places, the sea preserves better than land.

MB FACT: Living under pressure can literally change how you sound, which is how "underwater workers" sometimes feel like aliens to their own families. In the **North Sea oil fields**, saturation divers live for days or weeks in pressurized systems so they don't have to decompress after every shift. Breathing helium-rich gas can make voices turn high and squeaky -- real "chipmunk talk." It's funny until you realize it's a survival tool. Living below isn't just deeper -- it's a different version of being human.

MB FACT: Some of the most dramatic deep searches weren't for treasure -- they were for answers. When aircraft and major objects are lost over deep ocean, teams use **side-scan sonar** and deep robots to sweep huge areas like mowing a lawn, line by line. In wide open places like the **South Atlantic** or remote ocean basins, the search can feel endless -- just darkness and data until one strange shape appears. The moment a wreck outline pops onto the map is like finding a needle by scanning the entire planet's floor. Deep exploration is often built from stubborn, boring passes that suddenly turn into history.

MB FACT: Underwater habitats taught engineers to fear **tiny problems**. In **France's Conshelf program** in the **Mediterranean**, small leaks caused more danger than dramatic failures. Constant pressure slowly weakened materials without warning sounds or splashes. A pin-sized flaw could grow into a serious threat. Underwater living punished inattention more than bold mistakes.

MB FACT: Sonar alarms once lit up over what seemed like solid objects. During training exercises near **Japan**, naval operators detected dense shapes moving beneath the surface.

Later investigations revealed **massive schools of squid and fish**, packed so tightly they reflected sound like metal. Thousands of animals moving together looked like a single giant thing. Life fooled machines designed to find submarines.

MB FACT: Some of the loudest underwater sounds weren't machines at all. The **U.S. Navy's SOSUS listening network**, stretched across the **Atlantic and Pacific**, picked up deep, repeating pulses. Analysts first thought they were mechanical. They turned out to be **whales calling across entire ocean basins**, their voices traveling hundreds of miles. Military tools accidentally opened a new world of animal communication.

MB FACT: Maps of underwater caves refused to stay the same. Government divers exploring flooded **Mediterranean cave systems** found that passages **shifted shape between dives** as fine sediment moved. Sonar distances changed without warning. A tunnel that felt familiar could feel new days later. Underwater landscapes weren't fixed -- they were alive with motion.

MB FACT: Living underwater made people feel everything more intensely. At the **Aquarius Reef Base off Florida**, researchers noticed that **distant boats felt close**, and gentle vibrations traveled through walls and bodies. Sound carried farther and clearer underwater. Silence didn't exist. The ocean turned small movements into constant signals.

MB FACT: Rivers can hide entire fleets. In the **Yangtze River near Wuhan, China**, sonar surveys revealed dozens of **ancient shipwrecks buried under mud**, stacked from different centuries. Floods pushed boats into the same bends over and over again. From above, the river looks calm and ordinary. Underneath, it's a layered timeline of travel, trade, and disaster.

MB FACT: The seafloor can "swallow" wrecks without breaking them. In the **Black Sea**, explorers found ancient ships resting upright with **masts still standing**, even after **2,000+ years**. Deep water there has almost no oxygen, so wood-eating organisms can't survive. Ships don't rot -- they pause in time. Mapping the area feels like flying over frozen history.

MB FACT: Underwater volcanoes are mapped without ever seeing lava. Along the **Mid-Atlantic Ridge**, scientists use sonar to detect **fresh rock flows** by how sharply they reflect sound. Smooth new lava looks different than older, cracked rock. From sound alone, researchers can tell where Earth recently split open. The planet leaves clues -- even underwater.

MB FACT: Some shipwrecks are found by following cable damage. In parts of the **Mediterranean Sea**, engineers tracing broken **undersea communication cables** discovered wrecks snagging them. Heavy anchors and steel hulls dragged across the bottom like plows. Fixing the internet accidentally mapped the past. Modern tech ran straight into old history.

MB FACT: Ice can act like a giant underwater speaker. Beneath **Arctic sea ice**, submarines and research vehicles detect sounds traveling much farther than expected. The flat ice traps sound below, bouncing it sideways for miles. A distant crack can feel close. Under ice, quiet spreads instead of fading.

MB FACT: Sunken airplanes don't always fall apart. In **Lake Michigan**, cold water and low light preserved WWII-era aircraft with wings and markings still visible. Sonar images look like planes parked on a runway -- just underwater. Pilots trained nearby for decades without knowing what lay below. Freshwater can keep secrets surprisingly well.

MB FACT: Divers don't always trust their eyes -- even in clear water. In places like **Crater Lake, Oregon**, extreme clarity makes objects look closer and smaller than they are. A rock that seems within arm's reach might be several body lengths away. Depth messes with distance. Underwater vision lies politely.

MB FACT: Living underwater changed how people felt time pass. During the **SEALAB experiments off California**, aquanauts stayed submerged for weeks. Many reported that **days felt shorter and nights felt longer**, even with clocks nearby. Pressure, isolation, and constant noise reshaped sleep cycles. Humans could live below the surface -- but their brains didn't keep time the same way.

On a Lighter Side

MYTHS - BUSTED

MYTH: Lake Champlain was believed to hide long-living fish that predated settlement.

Early French settlers in the 1600s recorded local reports of massive fish that "had always been there." When oversized sturgeon were caught far larger than expected, people believed the lake preserved ancient life. The myth wasn't about monsters -- it was about time moving differently underwater.

MYTH: Some lakes "breathe" and will pull living things under.

In **Central Africa, in northern Cameroon near the border with Nigeria**, people believed certain crater lakes were alive and inhaled and exhaled. When animals or people drowned on calm days, locals said the lake had "breathed in." These waters were avoided at specific times of day and year. The myth wasn't about storms or waves -- it was about a lake that chose when to take things.

MYTH: One river in Europe could rearrange itself overnight.

Along the **Danube River in Central and Eastern Europe**, medieval boatmen believed the river moved its own sandbars while people slept. Safe routes marked one day vanished the next, even without floods. Without modern charts or markers, the river was said to "walk." The myth explained why the same river never behaved the same way twice.

MYTH: A lake in eastern Turkey followed different rules than all other water.

In **eastern Turkey, near the borders of Iran and Armenia**, travelers described Lake Van as unnatural. Its water tasted salty, yet fish lived inside it. People believed the lake rejected normal rules -- neither fresh nor sea, but something in between. The myth wasn't fear-based. It was confusion over a lake that didn't match expectations.

MYTH: A great African river disappeared underground and flowed back on itself.

For centuries, people believed the **Niger River in West Africa** vanished beneath the land before reappearing elsewhere. Its strange sideways curve across the continent didn't match how rivers were "supposed" to flow. Mapmakers repeated the belief well into the 1700s. The myth explained a river that refused to point toward the ocean.

MYTH: One lake in Central Asia was really two lakes sharing one body.

In **southeastern Kazakhstan, east of the Caspian Sea**, travelers believed Lake Balkhash was split in half. One side tasted fresh. The other tasted salty. Locals said the waters refused to mix, even though nothing separated them. Long before instruments confirmed the difference, the myth described a lake divided by an invisible line.

REAL LEGENDS

LEGENDS

The Silent Shapes of Loch Ness
Scotland -- Loch Ness

Long before modern sonar, British naval testing quietly took place in Scottish lochs because of their depth and isolation. Locals spoke of long, dark shapes moving steadily beneath the surface, leaving slow wakes without splashes. When sonar trials later mapped the loch, operators noted steep underwater slopes and deep trenches capable of hiding large moving objects from view until the last moment. The legend grew not from fantasy, but from repeated encounters with something that never surfaced fully. Even today, Loch Ness remains one of the most acoustically complex lakes in Europe, capable of producing movements that feel deliberate but resist explanation.

The Lights Beneath Lake Baikal
Siberia, Russia -- Lake Baikal

For decades, fishermen and researchers on Lake Baikal reported strange lights moving deep below the ice. These sightings often occurred during winter, when sound and light behave unusually under frozen surfaces. Soviet-era deep-water research and vehicle testing later revealed that submersibles and equipment were sometimes operating at depths few believed possible in a lake. Yet not all reports matched known activity. Baikal's immense depth, extreme clarity, and layered temperatures still create conditions where movement appears unnatural. Some encounters remain unaccounted for, even in declassified records.

The Moving Shadows of the Mediterranean
Mediterranean Sea -- Southern Europe

Ancient sailors described enormous shadows gliding beneath their ships, too large and smooth to be fish. Centuries later, modern navies conducting sonar mapping near underwater cliffs reported similar detections -- massive returns that moved, changed shape, and then vanished. Researchers eventually linked some of these to internal waves and biological layers riding steep terrain. But not every contact fit the pattern. The Mediterranean's crowded seafloor -- ancient ports, wrecks, and modern debris -- creates a layered mystery where history, geology, and life overlap. The legend persists because the sea itself still hides its motion.

The Watchers of the Underwater Caves
Mediterranean & Caribbean Regions

Divers mapping flooded cave systems for government research often report the same unsettling feeling: being followed. Lights illuminate movement just beyond reach, then darkness swallows it again. Sound bends unpredictably inside flooded tunnels, making direction unreliable. In some cases, divers have exited caves convinced something paced them silently. No creature has ever been confirmed -- but neither has the phenomenon been dismissed. These caves remain among the least understood underwater environments on Earth, where navigation tools and human instincts frequently fail together.

On a Lighter Side

DID YOU KNOW ?

Did You Know?

Much of the underwater world was mapped before scientists were even allowed to see it. During the Cold War, militaries quietly scanned lakes and seas to hide submarines, not to explore nature. Those maps later revealed ridges, sinkholes, and strange biological zones no one expected. Science didn't discover these places first. Survival did.

Did You Know?

Living underwater taught researchers something unsettling: **noise never stops below the surface**. Inside sealed habitats, pumps, air systems, and metal vibrations hum day and night. Crews slept, ate, and worked without silence. Over time, the sound wore them down more than danger ever did.

Did You Know?

Engineers often test underwater gear in lakes because calm water feels simpler than the ocean. No waves. No tides. But sound behaves strangely there. Layers of temperature bend sonar signals until echoes appear where nothing exists. Still water can lie.

Did You Know?

Some underwater "unknowns" were never solved because no one guessed wildly enough. Military analysts are trained to leave blanks when data doesn't repeat. If a contact appears once and never again, it stays unresolved. Mystery survives when certainty is resisted.

Did You Know?

Animals don't just live around underwater equipment -- they attack it. **Catfish** bite cables. **Crayfish** tear insulation. Mussels clog sensors until readings drift. Engineers learned the hard way that life below isn't passive. It pushes back.

Did You Know?

Maps of lake and river bottoms expire faster than road maps. Floods move sediment like avalanches. Algae and bacteria build layers that didn't exist months earlier. The ground below water is always shifting. Stability is an illusion.

Did You Know?

People living underwater lose their sense of time faster than their sense of fear. Without sunrise or sunset, days blur together. Crews rely on strict routines to stay mentally anchored. Below water, time must be manufactured.

Did You Know?

Some freshwater lakes preserve wrecks so well they look newly sunk. Cold, dark, oxygen-poor water slows decay to a crawl. Wood stays sharp. Metal holds shape. These lakes don't erase history -- they freeze it.

Did You Know?

Underwater caves confuse humans and machines in the same terrifying way. Sound ricochets, folds back on itself, and reverses direction. Navigation tools disagree. Even experts can't trust their senses. Below ground and underwater, space lies.

Did You Know?

Early underwater explorers learned a brutal lesson: depth rarely gives answers. Sonar returns fragments. Sensors contradict each other. Clear explanations are rare. The deeper humans go, the more patience matters.

STORY MOMENT

STORY MOMENT

The Map That Didn't Match the Water

The chart said the bottom was smooth.
Flat.
Safe.

That was the problem.

The research sub hovered quietly above the seafloor, its lights cutting thin cones through the dark as the operator adjusted the controls. On the sonar screen, the bottom looked boring -- just a gentle slope stretching away like an empty underwater plain. No sharp edges. No sudden drops. Nothing that demanded caution.

But the forward camera told a different story.

Out of the darkness, a shadow rose where the map insisted there was empty space. Not a bump. Not a hill. A wall.

They slowed instantly.

Years of training had taught the crew to trust their instruments -- but also to question them when something didn't feel right. The pilot nudged the sub forward, careful not to stir sediment that could blind them in seconds. The shadow sharpened into a steep underwater ridge, rough and broken, rising fast and close enough that a small drift could scrape metal against stone.

"Confirm position," someone said.

They did.
Then they checked again.

Same coordinates. Same depth. Same chart.

The map was wrong.

As the sub eased along the ridge, the sonar began to flicker. The clean outline stretched and warped, making the wall appear longer than it should have been. Almost like it was sliding as they watched. The cabin went quiet. No one joked. No one reached for the logbook.

No one said it out loud, but everyone felt it -- the uneasy sense that the seafloor wasn't frozen in place. It was still moving. Still reshaping itself beyond the reach of human eyes.

They backed away and marked the spot. No alarms. No emergency ascent. Just a calm note entered into the system: *terrain inconsistent with existing charts.*

Later, during the dive review, analysts pieced it together. The ridge wasn't new -- it was newly *moved*. A section of the underwater slope had collapsed years after the last survey, shifting tons of rock without a splash, without a sound anyone could hear above the surface.

Nothing dramatic had happened that day.
No one had been wrong.

The map was accurate... for a world that no longer existed.

That night, the crew talked less about the ridge itself and more about what it meant. Humans could live below. They could explore below. They could map the deep in incredible detail.

But the water didn't promise to stay still.

Down there, the planet could quietly rewrite itself -- long after the maps were printed.

FUN QUIZ

1. **Which underwater place preserved ancient ships so well that wooden masts were still standing?**
   A) Lake Michigan
   B) The Black Sea
   C) The Yangtze River
   D) Crater Lake

2. **Why can underwater maps near Japan become outdated very quickly?**
   A) Strong tides move the ships
   B) Fish schools block sonar
   C) Earthquakes shift the seafloor
   D) Ice melts and refreezes

3. **How were some Mediterranean shipwrecks discovered by accident?**
   A) Treasure hunters followed pirate maps
   B) Divers spotted bubbles from the wrecks
   C) Broken internet cables led to them
   D) Satellites photographed them from space

4. **What makes sound travel unusually far beneath Arctic sea ice?**
   A) Cold water slows sound down
   B) Ice absorbs sound completely
   C) Flat ice traps and reflects sound sideways
   D) Whales amplify the noise

5. **Why can objects in Crater Lake, Oregon, look closer than they really are?**
   A) The water bends light differently
   B) Fish create visual illusions
   C) Volcanic gases blur vision
   D) Ice reflects light downward

FUN QUIZ ANSWERS

1. B
2. C
3. C
4. C
5. A

CHAPTER 9

9-LIFE THAT SHOULDN'T EXIST...but DOES!

How is this possible? What Are They?

FUN & WEIRD FACTS

MB Fact = Mind-Blown Fact
Real discoveries. Real mysteries, Real Facts. Fun to make you think.

MB FACT: Way down in the **Izu—Bonin Trench near Japan**, there's a fish called the **hadal snailfish** that lives under pressure so intense it's like **an elephant standing on every inch of its body at the same time**. Sonar once spotted moving shapes that deep -- far below where fish were "allowed" to exist -- so yeah, people figured monsters made more sense than fish that strong.

MB FACT: Near underwater volcanoes on the **Galápagos Rift**, **giant tube worms** grow as tall as a grown-up... even though they don't have mouths, stomachs, or faces. When subs first shined lights on them, they looked like red-tipped towers just standing there, perfectly still, like something guarding the vents instead of eating near them.

MB FACT: Off **California in Monterey Canyon**, scientists kept seeing the same octopus sitting in the same spot for years. This deep-sea octopus guards her eggs without leaving to eat, slowly getting weaker the whole time. To sub pilots, it didn't look alive so much as *on duty*, like a silent watcher that never moved.

MB FACT: Near hot vents in the **Indian Ocean**, there's a snail that literally wears **iron armor**. The **scaly-foot snail** pulls metal from the water and builds it into its shell. When it was first spotted, it reflected light like steel, and people honestly wondered if it was part machine.

MB FACT: In the **Gulf of Mexico**, divers have reported sudden flashes underwater that look like something blinking back at them. These glowing bursts come from shrimp and fish that explode into light to confuse predators. To humans, it feels less like defense and more like, "Hey... something just noticed us."

MB FACT: Under almost **two miles of ice in Antarctica**, **Lake Vostok** has been sealed off from the surface for thousands of years. When radar picked up strange signals below the ice, people couldn't help wondering if something ancient had been trapped down there, living a life completely separate from the rest of Earth.

MB FACT: In the icy **Ross Sea near Antarctica**, **icefish** swim around with natural antifreeze in their blood. Early explorers watched fish moving under solid ice and realized the cold wasn't empty -- it was busy. That made polar waters feel a lot less safe and a lot more alive.

MB FACT: In the **Mariana Trench**, some animals can go years without eating. When food finally drifts down, the seafloor can suddenly fill with life. Sub crews said it looked like creatures were popping into existence, which is exactly how monster stories start.

MB FACT: Near **Sicily in the Mediterranean**, a tiny jellyfish called **Turritopsis dohrnii** can turn itself young again after getting hurt. Instead of dying, it resets. Sailors once claimed some jellyfish were impossible to kill -- and this one kind of proves they weren't imagining things.

MB FACT: Along the **Mid-Atlantic Ridge**, entire animal communities live around deep-sea vents with zero sunlight. Their food comes from bacteria powered by chemicals, not the sun. When explorers first saw full ecosystems glowing in total darkness, it felt like stumbling into a secret world that wasn't supposed to exist.

MB FACT: In frozen lakes across **northern Canada and Scandinavia**, some fish and insects survive months sealed under ice with almost no oxygen. Their bodies slow way down, barely moving at all. To people watching life disappear under solid ice, it looked like the lake had gone dead -- until things started moving again in spring.

MB FACT: Near deep-sea vents in the **Pacific Ocean off Costa Rica**, certain **marine worms** don't breathe oxygen at all. Instead, they absorb chemicals straight through their skin and use toxic mud as fuel. Early sub crews described these areas as "dead zones," right up until creatures started crawling out of what looked like poison.

MB FACT: In **Mono Lake, California**, tiny **brine shrimp and alkali flies** live in water so alkaline it would burn human skin. The lake looks calm and harmless, but almost nothing else can survive there. To early visitors, it felt like a place where normal rules stopped working.

MB FACT: In the deep waters off **Japan**, some fish have bones so soft they feel rubbery instead of hard. Solid bones would snap under pressure, so flexibility keeps them alive. When nets pulled them up, their strange texture made people think they weren't fully solid -- adding to the idea of deep-sea creatures built differently.

MB FACT: Near hydrothermal vents off **Easter Island**, **vent crabs** wave their claws through mineral-rich water to grow bacteria on them -- then eat the bacteria later. To early observers, it looked like the crabs were summoning food out of thin water, like underwater farming without farms.

MB FACT: In pitch-black caves and lakes like **Cavefish habitats in Mexico's Yucatán**, some fish lose their eyes completely, while others evolve super-sensitive ones. Darkness didn't erase vision -- it split it. To humans, eye-less fish felt especially unsettling, like something adapted to see in ways we couldn't.

MB FACT: In deep waters of the **Atlantic Ocean**, certain shrimp shoot out glowing clouds when threatened. The sudden flash scrambles a predator's senses long enough for the shrimp to escape. To divers, it looks like a creature vanishing inside its own light.

MB FACT: Beneath the seafloor off **South Africa**, scientists have found microbes living inside solid rock, fed by chemicals moving through tiny cracks. There's no open water, no sunlight, no space -- yet life is there. For years, unexplained signals from below the seabed fueled ideas that something was moving inside the Earth itself.

MB FACT: In deep lakes like **Lake Baikal in Siberia**, temperature layers trap animals at very specific depths. Swim too high or too low, and they can die. To fishermen, creatures seemed to appear and disappear without warning, as if the lake itself was rearranging where life was allowed.

MB FACT: Near vents in the **western Pacific**, some fish live right at the edge of scalding water hot enough to damage human skin. They constantly shift position to stay alive. To early sub pilots, it looked like animals dancing on invisible danger lines, surviving by perfect timing instead of strength.

MB FACT: In deep parts of the **Pacific Ocean**, animals like **sea cucumbers** can shut almost their entire bodies down when oxygen disappears. Organs pause, movement stops, and everything slows to near zero. To early divers, they looked dead -- until conditions improved and they quietly restarted.

MB FACT: In the deep Atlantic near **Newfoundland, Canada**, predators like **anglerfish and gulper eels** don't chase food at all. They hang perfectly still for hours or even days. When something finally wanders close, the attack is sudden. To sonar operators, these motionless shapes looked like lurking giants waiting to strike.

MB FACT: Near hydrothermal vents in the **western Pacific**, microbes like **Thermococcus** thrive in water hot enough to melt plastic. These heat-loving lifeforms don't tolerate warmth -- they *require* it. Early temperature readings made scientists think sensors were broken, because life wasn't supposed to exist there at all.

MB FACT: In the deep waters of **Monterey Bay, California**, a fish called the **barreleye** has a clear, see-through head with eyes that rotate inside it. Sub crews described it as a floating dome with moving eyes, something that didn't look like a fish -- or anything else familiar.

MB FACT: In isolated lakes like **Lake Baikal in Siberia**, animals such as the **Baikal seal** evolved completely cut off from the ocean. With no outside competition, life took strange turns. Creatures here don't match expectations, making the lake feel more like a biological experiment than a normal body of water.

MB FACT: In extreme environments like the **Arctic Ocean**, animals such as **polar cod** stack multiple survival tricks at once -- pressure tolerance, antifreeze blood, and toxin resistance. One adaptation isn't enough. To early explorers, these fish seemed nearly indestructible.

MB FACT: In deep oceans worldwide, fish like gulper eels and **viperfish** have jaws so loosely connected they can swallow prey bigger than themselves. Passing up food isn't an option when meals are rare. To sailors pulling up nets, these fish looked broken -- or monstrous.

MB FACT: Beneath thick winter ice on lakes like **Lake Superior**, sound travels farther and clearer than in open water. Animals such as **lake trout** rely more on vibration than sight. People hearing strange knocks and hums beneath the ice often blamed something large moving below.

MB FACT: In shallow seas around **coral reefs in the Caribbean**, bright red fish like **squirrelfish** look flashy in photos -- but underwater, red light disappears fast. That means red animals turn black and nearly invisible. To divers and sailors, something bright one moment and gone the next felt like a creature that could vanish on command.

MB FACT: Near hydrothermal vents along the **East Pacific Rise**, entire animal communities can appear and disappear within just a few years. When vents shut down, creatures like **tube**

worms, vent crabs, and vent shrimp die off fast. When new vents open elsewhere, life rushes in again. To early explorers, it felt like whole ecosystems were teleporting around the ocean floor.

MB FACT: Near vents along the **Juan de Fuca Ridge off the Pacific Northwest**, animals like **vent mussels** can sense tiny chemical changes in the water. They move before eruptions or collapses happen. To humans, it looks like animals reacting to danger before it even exists.

MB FACT: In oxygen-free zones of the **Black Sea**, creatures like **worms and clams** rely on internal bacteria to turn waste into energy. Nothing gets thrown away. To early researchers, life thriving in "dead zones" made the water feel unpredictable and alive in the wrong places.

MB FACT: Deep beneath the seafloor in places like the **Atlantic Ocean off South Africa**, microscopic life survives inside solid rock. Water carrying chemicals slips through tiny cracks, feeding bacteria hidden underground. For years, unexplained signals from beneath the seabed made people imagine things moving inside the Earth itself.

MB FACT: In the icy waters of **Alaska's lakes and rivers**, animals like **wood frogs and certain insects** can freeze solid during winter. Their hearts stop. Ice forms inside their bodies. When spring arrives, they thaw out and hop away like nothing happened. To people finding "dead" animals that later moved again, it felt impossible.

MB FACT: In deep waters of the **North Atlantic**, slow-moving animals like **greenland sharks** can live for centuries. They grow so slowly and face so few predators that time barely matters to them. Sailors pulling up ancient-looking sharks believed they were encountering creatures older than history itself.

MB FACT: In freshwater lakes like **Lake Tanganyika in Africa**, pressure increases differently than in the ocean because salt isn't involved. Fish adapted to deep freshwater can be crushed by sudden depth changes. To fishermen, healthy fish brought up from deep water sometimes died mysteriously, making the lake feel dangerous even when calm.

MB FACT:
The **Portuguese man o' war looks like a jellyfish** -- but it isn't one animal at all. It's a **floating team of tiny creatures**, each doing a different job, all stitched together into one dangerous package. Some parts sting, some parts digest food, and some just keep it afloat. Touching one is like getting attacked by an entire colony at once -- proof that teamwork can be terrifying.

MB FACT: In dark ocean waters near **Hawaii**, animals like **sea turtles and some fish** can sense Earth's magnetic field to navigate. They follow invisible lines humans can't feel. When creatures traveled long distances with perfect accuracy, it felt like they were being guided by something unseen.

MB FACT: In the deep ocean, many animals like **lanternfish and deep-sea squid** reproduce without ever meeting a mate. Eggs and sperm drift until chance brings them together. To humans, the idea of life continuing without contact made the deep feel lonely -- and strange.

MB FACT: In the deepest parts of the ocean, there is no day or night. Animals like **deep-sea worms and crustaceans** don't follow a daily schedule at all. They eat and move only when food arrives. To explorers, the deep felt timeless, like a place where clocks didn't work.

MB FACT: In deep-sea environments off **Japan and the Philippines**, some animals survive crushing pressure perfectly -- but die in shallow water. Creatures built for the deep aren't safer at the surface. To sailors, this flipped the idea of "safe waters" upside down.

MB FACT: In deep waters off **Norway and Greenland**, animals like **greenland sharks** move so slowly that algae can grow on their eyes. Sailors once mistook these drifting shapes for floating monsters, especially when they surfaced near boats without warning.

MB FACT: In underwater caves near **the Bahamas**, schools of fish can move so tightly together that sonar reads them as one giant object. Early operators thought they were tracking a single massive creature instead of thousands moving in perfect sync.

MB FACT: In cold lakes across **Russia and northern Europe**, trapped gas can suddenly release from the bottom, shaking ice and sending booming sounds across the surface. People blamed lake monsters long before learning the noises came from the lake itself shifting underneath them.

MB FACT: In murky rivers like the **Amazon**, animals such as **river dolphins and giant catfish** create huge rolling wakes just below the surface. At night, these moving bulges looked like single enormous creatures circling boats.

MB FACT: In deep ocean waters near **New Zealand**, giant squid arms have been found with suction scars larger than dinner plates. Sailors never saw the squid -- only the marks -- fueling stories of unseen beasts attacking from below.

On a Lighter Side

Look Mom, I grew my first tentacle!

"Well, that advertising was misleading...
He sure wasn't very crunchy."

MYTHS - BUSTED

MYTHS

MYTH: Life cannot exist without sunlight.

This idea comes from surface ecosystems where plants form the base of all food chains. For a long time, scientists assumed sunlight was a non-negotiable requirement for life anywhere. The discovery of deep-sea ecosystems powered entirely by chemical energy shattered that assumption. Life doesn't need light -- it needs energy, and chemistry can provide it.

MYTH: Extreme environments kill life instantly.

We tend to imagine poison, heat, pressure, or cold as absolute barriers. In reality, many organisms don't just tolerate extremes -- they depend on them. Remove the toxins, heat, or pressure, and these organisms die. "Extreme" is a human label, not a biological rule.

MYTH: Pressure always crushes living things.

High pressure feels dangerous because rigid objects collapse under it. But many deep-sea organisms aren't rigid at all. Their soft, flexible bodies distribute pressure evenly, preventing damage. Pressure becomes deadly only when structure fights it instead of yielding.

MYTH: Oxygen is required for all complex life.

Oxygen dominates surface biology, so it's easy to assume it's universal. Yet entire ecosystems operate in oxygen-poor or oxygen-free environments. Some organisms rely on chemical reactions and microbial partners instead. Oxygen is useful -- but it's not mandatory.

MYTH: Life must constantly move to survive.

Fast movement feels essential in predator-prey systems we see every day. In extreme underwater environments, movement wastes energy. Many creatures survive by waiting -- sometimes for months or years. Stillness is not weakness; it's strategy.

MYTH: Harsh environments produce small, weak organisms.

We often picture fragile life clinging barely to survival. In reality, some extreme environments produce surprisingly large or long-lived organisms. Reduced competition and specialized energy sources allow growth without constant threat. Harsh conditions can simplify life instead of shrinking it.

MYTH: If humans can't survive there, nothing can.

Human limits are often mistaken for biological limits. Our bodies are built for a narrow range of conditions. Life as a whole is not. What kills us may be perfectly normal to something else.

REAL LEGENDS

LEGENDS

The Men Beneath Lake Baikal
Lake Baikal, Siberia

During Cold War–era dives in Lake Baikal, Russian researchers reported encounters that were never fully explained. Divers described tall, humanoid shapes moving calmly in the water at extreme depths, without visible equipment. When a recovery team attempted to approach, several divers surfaced rapidly and suffered severe decompression injuries. Official explanations later blamed hallucinations caused by cold and nitrogen narcosis. Yet Baikal's depth, clarity, and ancient isolation continue to fuel questions. The lake holds more water than all North America's Great Lakes combined -- and much of it remains unexplored.

The Devil of the Deep Trench
Mariana Trench, Western Pacific

Early deep-sea probes sent into the Mariana Trench recorded sudden impacts and violent shaking, as if struck by something massive. Engineers assumed mechanical failure until repeated missions logged similar events. Some researchers speculated about pressure waves or collapsing terrain. Others quietly wondered if large, unknown animals were interacting with the equipment. No creature has ever been confirmed, but the trench remains deeper than Mount Everest is tall -- and mostly unseen.

The Living Waters of Lake Vostok
Antarctica

Lake Vostok lies buried beneath more than two miles of ice, sealed off for possibly millions of years. When scientists first proposed drilling into it, critics warned that nothing could survive there. Yet chemical signatures suggested biological activity. The idea that life could persist, isolated longer than humans have existed, felt almost unsettling. Even now, with limited sampling, researchers tread carefully. If life thrives there, it rewrites the timeline of survival itself.

The Glowing Sea of the Indian Ocean
Indian Ocean

Sailors for centuries reported vast stretches of ocean glowing softly at night, bright enough to read by. For generations, the phenomenon was dismissed as exaggeration or superstition. Modern satellite images eventually confirmed enormous glowing patches caused by bioluminescent bacteria, sometimes covering thousands of square miles. The scale shocked scientists. Entire oceans can light up without warning, turning darkness into something alive.

The Fish That Walked Out of the Abyss
Global deep-sea sightings

Throughout maritime history, sailors described grotesque, unfamiliar fish hauled up from extreme depths -- creatures with soft bodies, tiny eyes, or bizarre shapes. Many were discarded immediately, considered malformed or cursed. Only later did scientists realize these animals were perfectly adapted to deep water and deformed only when brought to the surface. Legends of "failed creatures" were really encounters with life that wasn't meant to leave the deep.

On a Lighter Side

"A PhD in Physics hey! Very impressive, but can you jump through hoops and balance a ball on your nose?"

Mermaids of the Dead Sea

DID YOU KNOW ?

Did You Know ? Scientists once believed extreme environments were biological dead ends. Today, they are often considered the most important places to study how life begins. Many researchers now believe early life on Earth may have started in conditions similar to deep-sea vents, not in calm surface waters. What looks hostile now may actually be anciently familiar to life.

Did You Know ? Some organisms living in extreme underwater environments change how their proteins fold so they don't break under pressure or heat. This isn't a small tweak -- it's a complete rewrite of cellular behavior. The same molecules that fail in shallow water remain stable miles below. Life doesn't just adapt on the outside; it adapts at the molecular level.

Did You Know ? Deep lakes and oceans preserve biological evidence better than land in many cases. Cold, dark, low-oxygen conditions slow decay dramatically. This allows scientists to study organisms, behaviors, and ecosystems that would vanish quickly at the surface. Depth acts like a natural time capsule.

Did You Know ? Some extreme-environment species appear unchanged for millions of years. Stable, isolated conditions reduce the need for rapid evolution. When an environment doesn't change, survival favors consistency over innovation. In the deep, ancient designs can outlast surface trends.

Did You Know ? Life around underwater volcanoes often grows faster than life in stable environments. These habitats are temporary, so organisms are built for speed rather than longevity. Grow fast, reproduce quickly, and move on. Evolution adapts not just to conditions, but to how long those conditions last.

Did You Know ? Discoveries of life in frozen or toxic waters on Earth directly influence where scientists search for life beyond our planet. Moons with icy shells and subsurface oceans are now considered top candidates for extraterrestrial life. What survives here expands what's possible everywhere.

Did You Know ? Many extreme underwater organisms rely on partnerships with microbes to survive. These relationships are so tight that neither partner can live alone. Survival becomes a team effort written into biology itself. Independence is not always an advantage.

MIND-BLOWN Cartoons

STORY MOMENT

The Camera That Kept Recording

The remotely operated vehicle wasn't supposed to stay down that long.

It was meant to be routine. Lower the camera. Scan the seafloor. Collect sediment images. Bring it back up. No surprises. No headlines. Just clean data to file away and move on from. The kind of mission crews half-joked about while checking cables.

But when the vehicle reached the bottom, the live feed didn't match the map.

The chart showed a smooth plain. What the camera saw was broken -- cracked rock, jagged ridges, folds in the seafloor that looked less like geology and more like something unsettled. Shadows moved where nothing was supposed to move. The landscape felt busy. Almost restless.

The pilot slowed the descent.

At first, the strange shapes looked like heat shimmer, the visual warping caused by chemical reactions leaking from the crust below. Then one of the shapes shifted -- against the current. Not drifting. Not tumbling. Moving with purpose.

The lights brightened.

A cluster of pale forms came into view, anchored to bare rock. They pulsed gently, opening and closing in a rhythm that looked uncomfortably like breathing. No sunlight reached this depth. No plants could survive here. No obvious food source drifted past. And yet there they were -- alive and settled in.

The control room went quiet.

Sensors spiked. Chemical readings jumped hard enough to trigger alerts. The vehicle's temperature warning flickered, dipped, then stabilized. The environment was hostile by every engineering standard. Pressure that could crush steel. Heat that warped alloys. Water so corrosive it ate at exposed surfaces.

But the organisms didn't care.

They hadn't adapted *to* the conditions. They had built their entire existence *around* them.

The vehicle hovered longer than planned.

Back on deck, engineers argued over the call. Pull it up now, or risk a few more minutes? Every second increased the chance of failure. The cables strained. Any sudden shift could mean losing the vehicle -- and the footage -- with it.

Still, the camera kept recording.

Below, nothing reacted. No scattering. No defense. No curiosity. The organisms didn't acknowledge the machine at all. It wasn't a threat. It wasn't food. It was irrelevant.

When the vehicle finally surfaced, hours later than scheduled, the footage spread fast. Scientists replayed it frame by frame. Life hadn't just survived there. It had settled in and flourished -- right where every assumption said it shouldn't exist.

Later, one researcher added a single line to the mission log:

"Conditions we consider extreme are only extreme from the outside. Down there, this is normal."

The camera didn't uncover monsters.

It uncovered something stranger.

Confidence.

FUN QUIZ

QUIZ

---

1. **What powers entire ecosystems around deep-sea hydrothermal vents?**

   A) Sunlight filtering through the water

   B) Chemical reactions using sulfur/methane

   C) Heat from the Moon's gravity

   D) Plant roots reaching down from land

2. **Why do many deep-sea animals handle crushing pressure so well?**

   A) They have extra-thick bones

   B) They fill their bodies with air

   C) Their bodies are soft and flexible, so pressure doesn't "crush" them the same way

   D) They constantly swim upward to relieve pressure

3. **Which creature is famous for thriving near hydrothermal vents despite having no mouth or stomach?**

   A) Giant tube worm

   B) Great white shark

   C) Sea turtle

   D) Manta ray

4. **True or False:** Sunlight is required for a food web to exist.

---

5. **Why are antifreeze proteins a big deal for polar fish and other cold-water life?**

A) They warm the water around the fish

B) They prevent ice crystals from damaging cells

C) They make blood thicker

D) They create oxygen from water

6. **Which is a common "deep survival" strategy when food can be rare for long periods?**

A) Constant fast movement

B) Eating every day at the same time

C) Waiting and conserving energy

D) Migrating to the surface every week

7. **Why does finding life in toxic, icy, or totally dark water matter to scientists looking beyond Earth?**

A) It proves aliens live in the ocean

B) It shows life can adapt to conditions we used to think were impossible

C) It makes submarines easier to design

D) It means the deep sea is fully mapped

FUN QUIZ ANSWERS

1. B
2. C
3. A
4. False
5. B
6. C
7. B

Composition of the Earth's Water

Where is Earth's Water?

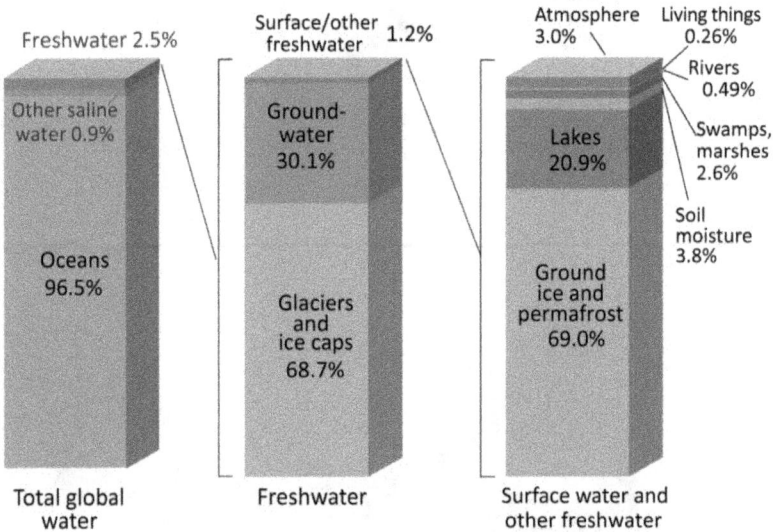

Freshwater 2.5%

Surface/other freshwater 1.2%

Atmosphere 3.0%

Living things 0.26%

Other saline water 0.9%

Ground-water 30.1%

Lakes 20.9%

Rivers 0.49%

Swamps, marshes 2.6%

Soil moisture 3.8%

Oceans 96.5%

Glaciers and ice caps 68.7%

Ground ice and permafrost 69.0%

Total global water

Freshwater

Surface water and other freshwater

Credit: U.S. Geological Survey, Water Science School. https://www.usgs.gov/special-topic/water-science-school
Data source: Igor Shiklomanov's chapter "World fresh water resources" in Peter H. Gleick (editor), 1993, Water in Crisis:
A Guide to the World's Fresh Water Resources. (Numbers are rounded).

CHAPTER 10

10-NAMES YOU'VE HEARD – FROM DEEP BELOW

Where Ships and Planes Were Lost at Sea

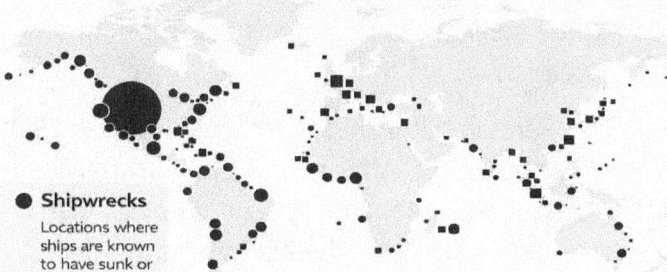

● Shipwrecks

Locations where ships are known to have sunk or been lost in water, including oceans, seas, and major coastal routes.

■ Aircraft Lost at Sea

Areas where airplanes are believed or confirmed to have gone down in water. (Some locations are based on last known position, not recovered)

● Circle Size

● Larger circles = more recorded losses in that area

● Smaller circles = fewer recorded

⚠ Important:

This map shows **patterns**, not exact counts.

Busy routes naturally have more recorded losses.

This map shows known locations where ships sank and airplanes were lost in oceans and large bodies of water over many centuries, from ancient sailing days to modern times. Each circle marks **an area**, not a single **wreck**. Bigger circles mean **more recorded losses** in that region, often because many people traveled the same routes.

Not every loss was caused by **mystery**. Most happened because of storms, war, navigation mistakes, or busy shipping lanes. Some wrecks are still unexplored, and many more were never recorded at all.

This map shows known locations where ships sank and airplanes were lost in oceans and large bodies of water over many centuries, from ancient sailing days to modern times. Each circle marks an area, not a **single wreck**. Bigger circles mean more recorded losses in that region, often because many people traveled the same routes.

FUN & WEIRD FACTS

MB Fact = Mind-Blown Fact
Real discoveries. Real mysteries, Real Facts. Fun to make you think.

MB Fact: The **Titanic** still rests nearly 13,000 feet below the North Atlantic, split into two massive sections. The bow looks eerily intact, while the stern is twisted and shredded from its violent descent. Cold, dark water slowed decay for decades, preserving dishes, shoes, and railings. Even now, bacteria are slowly eating the steel itself. The ship is not frozen in time -- it is quietly disappearing.

MB Fact: The German battleship **Bismarck** lies over 15,000 feet deep in the Atlantic Ocean. After one of the most famous naval chases in history, it was intentionally scuttled by its own crew. When explorers finally found it, the hull was largely intact. The damage that sank it came mostly from within, not enemy shells.

MB Fact: The **USS Arizona**, sunk during the attack on Pearl Harbor, still leaks small amounts of oil today. Sailors call it "the tears of the Arizona." The ship is considered a war grave, and parts remain sealed forever. It is both a wreck and a memorial.

MB Fact: The Swedish warship **Vasa** sank on its very first voyage in 1628 after a gust of wind tipped it over. It sat preserved in cold Baltic water for more than 300 years. When raised, it was shockingly intact, complete with carvings and cannons. The water had protected it better than any museum could.

MB Fact: The **Endurance**, Ernest Shackleton's Antarctic expedition ship, was crushed by ice in 1915. For over a century, no one knew exactly where it lay. When found in 2022, it appeared almost perfectly preserved. The cold, dark water of the Weddell Sea had guarded it like a vault.

MB Fact: The passenger ship **Lusitania** was sunk by a German submarine in 1915, helping pull the United States toward World War I. It lies off the coast of Ireland. The wreck sparked decades of debate over secret cargo aboard. Even now, questions remain about what was truly inside.

MB Fact: Thousands of **World War II aircraft** rest underwater after training accidents. Many pilots practiced takeoffs and landings over lakes and coastal waters. When planes crashed, they often sank quickly and were never recovered. Some still sit upright on the lake floor, almost ready to fly again.

MB Fact: The **Japanese aircraft carrier Akagi** lies deep beneath the Pacific near Midway. It was scuttled after being badly damaged in battle. The wreck helped confirm historical battle accounts decades later. The ocean preserved the truth when paperwork did not.

MB Fact: In Lake Michigan, dozens of **WWII-era planes** rest on the bottom from pilot training missions. Cold freshwater slowed corrosion dramatically. Propellers, wings, and even paint markings remain visible. The lake became an accidental museum.

MB Fact: The **Andrea Doria**, an Italian luxury liner, sank after colliding with another ship in 1956. It lies off the U.S. East Coast and is considered one of the most dangerous wreck dives in the world. Strong currents and collapsing interiors make exploration risky. The ship continues to claim lives long after sinking.

MB Fact: Ancient **Roman cargo ships** have been found with amphorae still stacked neatly inside. These clay containers once held olive oil, wine, and grain. Some wrecks date back over 2,000 years. The sea preserved everyday life better than land ever could.

MB Fact: The Greek warship **Antikythera wreck** revealed one of the most mysterious objects ever found underwater. Among statues and cargo was a complex gear system now known as the Antikythera Mechanism. It functioned like an ancient computer. Its discovery rewrote what we thought ancient technology could do.

The Endurance, Ernest Shackleton's Antarctic expedition ship

MB Fact: The **Black Swan Project** uncovered a massive treasure of silver coins from a Spanish ship lost in the 1800s. The legal battle over who owned it lasted years. The wreck proved that the ocean does not forget wealth. It just hides it.

MB Fact: The **Mary Rose**, flagship of England's King Henry VIII, sank during battle in 1545. When raised, it revealed weapons, tools, and personal items from the crew. Shoes still fit the shape of the men who wore them. History surfaced in human scale.

MB Fact: The legendary **Flying Dutchman** inspired stories of ghost ships, but real abandoned vessels drifting at sea fueled the myth. Storms, disease, and accidents sometimes left ships empty. Finding a ship with no crew created terror and imagination. Legends grew from real fear.

MB Fact: In the Black Sea, ancient ships remain astonishingly intact because deep water lacks oxygen. Wooden masts still stand upright. Rope and hull planks survive for centuries. It's one of the best ship graveyards on Earth.

MB Fact: The **HMS Hood**, once the pride of the British navy, exploded and sank in minutes during WWII. Over 1,400 sailors were lost. The wreck was discovered decades later in two main pieces. It confirmed how sudden and catastrophic the loss had been.

MB Fact: The **USS Indianapolis** delivered parts of the first atomic bomb before being torpedoed in 1945. Survivors spent days in the water surrounded by sharks. The wreck was found in 2017. Its discovery finally brought closure to families.

MB Fact: The **Spanish treasure fleet of 1715** was destroyed by a powerful hurricane off the coast of **Florida**. Eleven ships sank in a single night, scattering gold, silver, and jewels across the seafloor. Some treasure has been recovered over the centuries, often after storms shift the sand. But large portions remain buried or unreachable. Every hurricane season still has the power to reveal pieces of a 300-year-old disaster.

MB Fact: The aircraft carrier **USS Yorktown**, sunk during the **Battle of Midway in 1942**, lies more than **16,000 feet deep in the Pacific Ocean**. When explorers found it decades later, the ship's name was still readable on its hull. The wreck confirmed battle damage patterns long debated by historians. The ocean kept the evidence intact.

MB Fact: In **1944**, the German submarine **U-576** was sunk off the coast of **North Carolina** during World War II. It now rests near the wreck of a merchant ship it attacked moments earlier. Both vessels lie close together on the seabed, frozen in combat. It's one of the rare underwater battlefields preserved exactly where it happened.

MB Fact: The luxury liner **SS Central America** sank in **1857** during a hurricane in the Atlantic while carrying tons of California Gold Rush treasure. The ship went down off the coast of **South Carolina**. Over 400 people died. When the wreck was found in 1988, it was nicknamed "The Ship of Gold." Much of its cargo had never been touched.

MB Fact: The Japanese battleship **Musashi**, one of the largest warships ever built, was sunk in **1944** during the Battle of Leyte Gulf. It was discovered in **2015** in the **Sibuyan Sea, Philippines**, at extreme depth. The wreck showed massive torpedo damage. Even the largest machines humans built could vanish beneath the sea.

MB Fact: The Swedish warship **Mars**, sunk in **1564** during a naval battle in the Baltic Sea, was rediscovered in **2011**. Cold, low-oxygen water preserved cannons, weapons, and the hull remarkably well. The ship had exploded after catching fire. Nearly 450 years later, the sea still held its story.

MB Fact: In **Lake Toplitz** in **Austria**, crates of Nazi counterfeit money were dumped near the end of World War II. The cold, dark lake preserved the boxes on the bottom. Thick layers of fallen trees made exploration dangerous. The lake became known as one of Europe's most mysterious underwater sites.

MB Fact: The **HMS Victory**, a different ship than the famous one on display in England, sank in **1744** in the English Channel during a storm. Over 1,000 sailors were lost. The wreck was found in **2008**, scattered across the seabed. Bronze cannons confirmed its identity centuries later.

MB Fact: The ancient city of **Heracleion**, also called Thonis, sank near the **Nile Delta** around **1,200 years ago**. Earthquakes and rising water slowly swallowed it. When rediscovered in the early 2000s, temples, statues, and ships were found underwater. A major port had simply slipped beneath the sea.

MB Fact: During World War II, **dozens of aircraft** crashed into **Lake Erie** during training missions. Cold freshwater preserved many of them better than ocean wrecks. Some planes were later recovered almost intact. The lake quietly stored pieces of wartime history.

MB Fact: The **USS Lexington**, sunk in **1942** during the Battle of the Coral Sea, was found in **2018** more than **10,000 feet deep**. Aircraft still sat on its deck. Paint markings were visible decades later. Depth became protection.

MB Fact: The legendary **Antikythera shipwreck**, dating to around **100 BCE**, lies off the coast of **Greece**. Among statues and cargo, divers recovered a complex mechanical device. Today it's considered the world's oldest known analog computer. Ancient sailors carried technology far ahead of its time.

MB Fact: The **German High Seas Fleet** was scuttled by its own crews in **1919** at **Scapa Flow, Scotland**. Over 50 ships were deliberately sunk. Many were later salvaged, but some remain on the seabed. The ocean became a graveyard by choice, not battle.

MB Fact: The **SS Edmund Fitzgerald** sank in **1975** in **Lake Superior** during a fierce storm. All 29 crew members were lost. The cold freshwater preserved the wreck in haunting condition. It remains one of the Great Lakes' most famous tragedies.

MB Fact: In **1945**, Japanese submarines known as **I-400 class** were captured by the U.S. Navy. Rather than risk their technology being studied by rivals, the submarines were deliberately sunk near **Hawaii**. Their wrecks were rediscovered decades later. Even secrets can end up underwater.

MB Fact: The French passenger ship **Empress of Ireland** sank in **1914** after colliding with another vessel in the **St. Lawrence River, Canada**. It went down in just 14 minutes, faster than the

Titanic. Over 1,000 people died. The wreck still lies in the river, quietly marking one of the worst maritime disasters most people have never heard of.

MB Fact: In **1943**, the British battleship **HMS Royal Oak** was sunk at anchor in **Scapa Flow, Scotland** by a German submarine. It was supposed to be one of the safest naval harbors in the world. Over 800 sailors were lost. The wreck remains a protected war grave beneath calm waters.

MB Fact: The ancient port city of **Pavlopetri**, off the coast of **Greece**, sank more than **5,000 years ago**. Streets, buildings, and courtyards are still visible underwater. Unlike legendary cities, this one is unquestionably real. People once walked those roads, and now fish swim above them.

MB Fact: The American aircraft carrier **USS Hornet**, famous for launching the Doolittle Raid, was sunk in **1942** during World War II near the **Solomon Islands**. When found in **2019**, the ship's nameplate was still readable. Aircraft were discovered nearby. The ocean preserved history better than paper records.

MB Fact: In **Lake Mead**, Nevada, old towns were flooded when the Hoover Dam was completed in the **1930s**. Roads, buildings, and foundations still sit underwater. When water levels drop, pieces reappear. The lake hides an entire vanished landscape.

MB Fact: The **SS Thistlegorm**, a British supply ship sunk in **1941** in the **Red Sea**, still contains motorcycles, trucks, boots, and rifles. It was bombed during World War II while anchored. Divers today can still see the cargo frozen in time. It's like walking through a sunken warehouse.

MB Fact: The Spanish warship **Nuestra Señora de Atocha** sank in **1622** during a hurricane near the **Florida Keys**. It carried enormous amounts of silver, gold, and emeralds. Parts of the treasure were recovered centuries later, but much remains missing. Storms still shift the seabed, occasionally revealing clues.

MB Fact: During World War II, the Japanese sent **balloon bombs** across the Pacific toward North America. Some fell harmlessly, but others ended up in lakes and coastal waters. Pieces are still occasionally discovered underwater. Even weapons can vanish into silence.

MB Fact: The **SS Yongala**, sunk in **1911** during a cyclone off **Queensland, Australia**, lies in warm tropical water. Over time, coral and fish covered the wreck completely. What was once a tragedy became a living reef. Nature repurposed disaster.

MB Fact: The ancient lighthouse city of **Pharos**, once one of the Seven Wonders of the Ancient World, collapsed into the sea near **Alexandria, Egypt** around **1,300 years ago**. Massive stone blocks and statues still rest underwater. One of history's greatest landmarks didn't disappear -- it sank.

MB Fact: In **Lake Superior**, the wreck of the **SS Kamloops**, sunk in **1927**, contains a preserved crew member often called "Old Whitey." Cold freshwater slowed decay dramatically. The wreck turned into a chilling time capsule. Lakes can preserve just as powerfully as oceans.

MB Fact: The German submarine **U-85**, sunk in **1942** off the coast of **North Carolina**, was discovered decades later with its deck gun still visible. It sits within an area known as the Graveyard of the Atlantic. Hundreds of ships were lost there due to storms and war. The seabed is crowded with stories.

MB Fact: The luxury liner **SS America**, renamed multiple times, ran aground in **1994** near the **Canary Islands** during a storm. Over years, waves broke the ship apart. People watched it disappear piece by piece. Even modern ships can be claimed quickly.

MB Fact: The **Battle of Bikini Atoll** in **1946** left dozens of ships deliberately sunk during nuclear testing. Aircraft carriers, battleships, and submarines rest together on the seafloor. Radiation forced humans away, but marine life returned in huge numbers. The ocean adapted faster than expected.

MB Fact: The ancient city of **Atlit Yam**, off the coast of **Israel**, dates back nearly **9,000 years**. Stone houses, wells, and graves sit underwater. Rising seas swallowed the settlement slowly. It may be one of the earliest drowned cities known.

MB Fact: The American submarine **USS Tang**, one of the most successful subs of World War II, sank in **1944** in the **Taiwan Strait** after its own torpedo malfunctioned and circled back. Survivors escaped using experimental breathing gear. The wreck was located decades later. It's a reminder that underwater warfare was dangerous even without enemies nearby.

MB Fact: The German submarine **U-166** was sunk in **1942** in the **Gulf of Mexico**, just miles from the U.S. coast. For years, people believed German subs never reached that far inland. The discovery of the wreck proved otherwise. The war had crept much closer to home than anyone realized.

MB Fact: During World War II, dozens of **B-29 Superfortress bombers** crashed into the **Pacific Ocean** during long missions from island bases. Mechanical failures were common. Many of these aircraft were never found. Somewhere below, giant wings still rest in darkness.

MB Fact: The Japanese submarine **I-52**, sometimes called the "Golden Submarine," was sunk in **1944** in the **Atlantic Ocean** while carrying valuable cargo to Germany. It lay undiscovered for decades. When found, it sparked debate over what it still carried. Treasure rumors followed the wreck into modern times.

MB Fact: The Soviet submarine **K-129** sank in **1968** in the **Pacific Ocean** under mysterious circumstances. The U.S. secretly attempted to raise part of it during the Cold War. The operation was so secret that many details remain classified. The wreck sits deep, still holding unanswered questions.

MB Fact: The German submarine **U-352** was sunk in **1942** off the coast of **North Carolina**. Survivors were rescued, and the wreck later became a popular dive site. Its conning tower still stands upright. The Atlantic coast hides a dense underwater battlefield.

MB Fact: During the **Battle of Midway in 1942**, dozens of aircraft crashed into the sea during combat and emergency landings. Pilots sometimes ditched their planes intentionally. Many aircraft remain undiscovered. The ocean absorbed the chaos of battle.

MB Fact: The British submarine **HMS Perseus** sank in **1941** near the coast of **Greece** after hitting a mine. One crew member escaped using a special underwater breathing device. His survival was doubted for years. Later discoveries proved his story true.

MB Fact: In **1946**, the U.S. Navy deliberately sank tanks, landing craft, and vehicles during nuclear tests at **Bikini Atoll**. Armored vehicles now rest beside ships on the seafloor. Coral has grown over gun barrels and treads. War machines became reefs.

MB Fact: The Japanese aircraft **Mitsubishi A6M Zero**, a famous WWII fighter, crashed into **Lake Taal in the Philippines** during training. Freshwater preserved parts of the plane. Discoveries like this help historians confirm production details. Even lakes can hold aviation history.

MB Fact: The American submarine **USS Scorpion** sank in **1968** in the **Atlantic Ocean** under unknown circumstances. It rests nearly **10,000 feet deep**. The cause of its sinking has never been officially confirmed. The mystery still lingers.

MB Fact: In **Lake Ontario**, World War II training aircraft from Canada's British Commonwealth Air Training Plan rest on the bottom. Thousands of pilots trained there. Not every flight ended safely. The lake quietly remembers.

MB Fact: The German **Panzer tanks** sunk during river crossings in Eastern Europe during WWII are still occasionally found underwater. Rivers became unexpected battlefields. Vehicles meant for land ended up buried in mud. Water claimed them without warning.

MB Fact: The Soviet submarine **K-278 Komsomolets** sank in **1989** in the **Norwegian Sea** after a fire onboard. It reached extreme depths before sinking. Concerns remain about nuclear materials inside. The wreck is closely monitored even today.

MB Fact: In **Lake Baikal, Russia**, the world's deepest lake, entire research vehicles have been lost to depth and ice. Some sit more than **5,000 feet down**. The lake is so old and cold that metal corrodes slowly. What sinks there can remain untouched for decades.

MB Fact: Lake Superior, bordering the United States and Canada, holds over **6,000 known shipwrecks**. Sudden storms can rise without warning, even today. Cold freshwater preserves hulls, tools, and sometimes cargo. The lake earned the nickname "the inland sea that never gives up its dead."

MB Fact: In **Lake Champlain**, between **New York and Vermont**, wrecks from the **American Revolutionary War** still rest on the bottom. Wooden gunboats and cannons have been found nearly intact. Freshwater slowed decay dramatically. A battlefield became a time capsule.

MB Fact: The **St. Lawrence River** in Canada hides wrecks from more than **400 years of shipping**. Strong currents pulled ships under during storms and collisions. Some wrecks are stacked almost on top of each other. The river became a layered history book.

MB Fact: In **Lake Garda, Italy**, Mussolini's World War II military boats were deliberately sunk near the war's end. The vessels were hidden to keep them from enemy forces. They still rest in deep freshwater. Politics vanished, but hardware remained.

MB Fact: The **Danube River**, flowing through ten European countries, contains sunken tanks, bridges, and military vehicles from World War II. Retreating armies destroyed crossings to slow enemies. Some vehicles slipped into the river and were never recovered. Europe's longest river quietly swallowed war.

MB Fact: In **Lake Michigan**, the remains of **wooden schooners from the 1800s** sit upright on the lakebed. Cargo like coal and lumber still lies inside. Cold water preserved them so well that divers say it feels like stepping back in time. The lake acts like a freezer.

MB Fact: The **Volga River**, Russia's longest river, hides war debris from the **Battle of Stalingrad (1942–1943)**. Boats, bridges, and equipment sank during fighting. Mud and silt buried much of it. History disappeared beneath flowing water.

MB Fact: In **Lake Toplitz, Austria**, deep cold water preserved sunken crates dumped by Nazi forces in **1945**. The lake floor is littered with fallen trees, making dives dangerous. Some crates contained counterfeit money. Others may still be undiscovered.

MB Fact: The **Hudson River**, New York, holds Revolutionary War ships deliberately sunk to block British forces. These ships formed underwater barriers called chevaux-de-frise. Iron spikes were attached to wooden frames. Defensive structures still lie beneath the river's surface.

MB Fact: The **Mississippi River** hides steamboats that exploded, burned, or sank during the 1800s. Shifting currents buried many beneath layers of mud. Entire ships can vanish without leaving surface clues. The river constantly rewrites its own history.

MB Fact: In **Lake Mälaren, Sweden**, Viking-era ships and medieval boats were lost during trade and travel. Some wrecks still contain tools and cargo. Low oxygen levels helped preserve wood. Everyday life from a thousand years ago still waits underwater.

MB Fact: The **Rhine River**, flowing through central Europe, contains Roman bridges and military crossings destroyed over centuries. Stone foundations remain anchored to the riverbed. Water erased the surface, but not the past. Ancient engineering still shapes the river floor.

MB Fact: In **Lake Ontario**, Canadian and British training aircraft from World War II rest on the bottom. Many pilots survived, but the planes sank quickly. Cold water preserved control panels and frames. The lake became an unplanned aviation archive.

MB Fact: The American aircraft **Grumman F4F Wildcat** was a key fighter in early World War II. Several Wildcats crashed into **Lake Michigan** during pilot training in the 1940s. Cold freshwater preserved them almost perfectly. Some were later raised and restored to flying condition.

In Honor of those Lost at Sea
Author: Unknown

IN WATERS DEEP

In ocean waves no poppies blow,

No crosses stand in ordered row,

There young hearts sleep... beneath the wave...

The spirited, the good, the eternally brave,

But stars a constant vigil keep,

For them who lie beneath the deep.

'Tis true you cannot kneel in prayer

On certain spot and think. "He's there."

But you can to the ocean go...

See whitecaps marching row on row;

Know one for him will always ride...

In and out... with every tide.

And when your span of life is passed,

He'll meet you at the "Captain's Mast."

And they who mourn on distant shore

For sailors who'll come home no more,

Can dry their tears and pray for these

Who rest beneath the heaving seas...

For stars that shine and winds that blow

And whitecaps marching row on row.

And they can never lonely be

For when they lived... they chose the sea.

May God rest their souls

MYTHS - BUSTED

MYTH: The Bermuda Triangle causes ships and planes to vanish for unknown reasons.

People believed this because of reports of lost vessels and strange compass readings. In reality, the area is heavily traveled and has powerful storms, shifting currents, and busy shipping lanes. Navigation errors explain many incidents. There is no evidence of supernatural forces at work.

MYTH: The Kraken was completely imaginary.

Sailors described giant tentacled creatures rising from the sea. Today, scientists know giant and colossal squid exist and can grow to enormous sizes. Rare surface sightings likely inspired the stories. The monster was exaggerated, but the animals were real.

MYTH: Lake monsters like Nessie were invented stories.

Sightings in lakes have been reported for centuries. Science shows that waves, floating debris, large fish, and even swimming animals can create convincing illusions. No unknown monster species has been found. The sightings came from real observations, misinterpreted.

MYTH: Ancient sailors believed the ocean ended at a waterfall.

This idea comes from misunderstandings of early maps and myths. Sailors knew the sea continued beyond the horizon long before global maps existed. Ocean travel expanded steadily over centuries. Fear of the unknown, not belief in waterfalls, shaped early stories.

MYTH: Ghost ships really sail on their own centuries after sinking.

Abandoned ships drifting at sea terrified sailors. In reality, storms, currents, and failed evacuations explain empty vessels.

Ships can drift for long distances without crews. No supernatural forces are required.

MYTH: Submarines disappear more often than governments admit.

Cold War secrecy fueled this belief. While some losses were classified, most submarine sinkings are now documented. A few causes remain unknown due to depth and damage. Mystery does not equal conspiracy.

MYTH: Sunken treasure ships are cursed.

Accidents and deaths during salvage attempts led to this belief. In truth, dangerous diving conditions cause injuries and fatalities. Deep water, poor visibility, and unstable wrecks are the real risks. There is no curse -- only physics.

MYTH: The ocean floor is flat and empty.

Early explorers had no way to see deep underwater. Modern mapping shows mountains, valleys, volcanoes, and trenches. Entire ecosystems thrive on the seafloor. It is one of the most complex landscapes on Earth.

MYTH: Rivers can swallow ships whole without warning.

Stories came from ships vanishing quickly in muddy water. Strong currents, shifting sandbars, and sudden floods can trap vessels. Ships sink through natural forces, not sudden disappearances. Rivers are dynamic, not magical.

MYTH: The sea keeps whatever it takes forever.

Sailors believed nothing returned from the deep. Cold freshwater and deep oceans often preserve wrecks extremely well. Many ships are rediscovered centuries later. The sea hides things -- but not forever.

MYTH: Underwater cities exist only in legends.

Atlantis popularized the idea. Archaeology has now confirmed multiple real sunken cities worldwide. Earthquakes, flooding, and rising seas caused them to sink. Legends preserved memories of real disasters.

MYTH: Ancient people couldn't build long-lasting structures near water.

This belief came from missing ruins. Underwater discoveries reveal harbors, roads, and buildings made with advanced engineering. Many were lost due to natural changes. Ancient builders were far more skilled than once thought.

MYTH: Storms at sea are random and unexplainable.

Sailors once blamed angry gods. Today, meteorology explains storms through pressure systems and temperature changes. Storms follow patterns, even if dangerous. Knowledge replaced fear.

MYTH: Whirlpools drag entire ships straight down.

Stories exaggerated real whirlpools. Most whirlpools are not powerful enough to sink large ships. Dangerous currents exist, but ships are not sucked under whole. The myth grew from dramatic storytelling.

MYTH: Lakes never change once they fill.

People see lakes as calm and stable. In reality, water levels rise and fall, shorelines shift, and towns have been flooded. Human activity and climate drive change. Lakes are constantly evolving.

MYTH: Naval battles leave nothing behind underwater.

Explosions suggested total destruction. In fact, wrecks, weapons, and debris often remain intact. Entire battlefields still exist underwater. The sea preserves moments of conflict.

MYTH: Ancient maps showing sunken lands were pure fantasy.

Some maps were inaccurate, but others reflected coastlines before flooding. Rising seas reshaped continents after the Ice Age. Maps captured older geography. They weren't all imaginary.

MYTH: Planes that crash into water are destroyed instantly.

Hollywood reinforced this idea. Many aircraft ditch gently or sink intact. Cold water preserves them for decades. Some are later recovered almost complete.

MYTH: The deepest ocean is virtually lifeless.

This belief lasted until the 20th century. Scientists discovered life thriving even in total darkness. Some organisms rely on chemicals instead of sunlight. Life adapts in extreme ways.

MYTH: Fish and sea creatures avoid shipwrecks.

People assumed wrecks were dead zones. In reality, wrecks attract marine life. They provide shelter and food. Many become thriving reefs.

MYTH: Strange underwater sounds come from sea monsters.

Unexplained noises frightened sailors. Scientists later identified ice movement, earthquakes, animals, and human activity as sources. The sounds were real, but monsters were not. Nature is noisy.

MYTH: Sea serpents were just giant snakes.

Early sightings didn't match snakes exactly. Oarfish, eels, and whales can appear serpent-like at the surface. Poor visibility stretched the truth. Real animals inspired the stories.

MYTH: Pirates always buried treasure on land.

Stories focused on maps and islands. In reality, pirates often lost treasure at sea. Shipwrecks scattered cargo underwater. Most pirate treasure was never buried neatly.

MYTH: Cold water destroys wrecks faster than warm water.

People assumed cold was harsher. Cold water actually slows corrosion and decay. Many freshwater wrecks remain remarkably intact. Warm water often causes more damage.

MYTH: Ancient floods described in myths never happened.

Flood stories exist worldwide. Geological evidence confirms massive flooding events in the past. Myths preserved memories before written records. Stories survived when facts were forgotten.

MYTH: Large lakes don't have tides or currents.

They lack ocean tides, but internal currents exist. Wind and temperature differences move water constantly. Lakes behave more like seas than ponds. Motion shapes everything beneath them.

MYTH: All famous wrecks have been found.

Media coverage gives that impression. Many famous ships, planes, and submarines remain undiscovered. New finds happen every year. The map is incomplete.

MYTH: Sunken warships are empty shells.

People imagine stripped metal. Many still contain equipment, personal items, and sealed rooms. Some are protected graves. They remain powerful historical sites.

MYTH: Water erases history faster than land.

Time and erosion affect land heavily. Underwater conditions can slow change dramatically. Some history survives better underwater. Water sometimes protects rather than destroys.

MYTH: If something sinks deep enough, it can never be seen again.

This was once true. Modern sonar, robots, and mapping changed that. Deep discoveries now happen regularly. Depth is no longer absolute hiding.

LEGENDS

REAL LEGENDS

ATLANTIS -- The City That Slipped Beneath the Sea
(Atlantic Ocean / Mediterranean region)
Plato wrote of a powerful island city swallowed by the sea in a single night. For centuries, people argued whether Atlantis was real or just a story. Earthquakes, volcanic eruptions, and sudden floods have destroyed real coastal cities throughout history. Some believe Atlantis may be a memory of one of these disasters, retold until it became legend.

THE KRAKEN -- The Monster That Dragged Ships Under
(North Atlantic, near Norway and Greenland)
Sailors spoke of something massive rising beneath their ships, pulling them down with tentacles. These stories terrified crews for generations. Today, we know giant and colossal squid live in deep waters and can grow to enormous sizes. Rare surface encounters may have turned real animals into nightmare legends.

THE FLYING DUTCHMAN -- The Ship That Never Reaches Shore
(Southern oceans near Cape of Good Hope)
The legend tells of a cursed ship doomed to sail forever, never docking. Sailors claimed seeing glowing ships in storms meant disaster was near. Scientists now know mirages can lift distant ships into the sky, making them appear ghostly. Fear and exhaustion filled in the rest.

THE LOST FLEET OF KUBLAI KHAN

(Sea of Japan)

The Mongol emperor Kublai Khan sent massive fleets to invade Japan in the 1200s. Both times, powerful storms destroyed his ships. The Japanese called the storms *kamikaze*, or "divine wind." Shipwrecks have since been found underwater, proving the fleets were real -- and truly lost.

THE MARY CELESTE -- The Ship with No One On Board

(Atlantic Ocean, near the Azores)

In 1872, a ship was found drifting with no crew aboard. Food, cargo, and personal belongings were untouched. There were no signs of struggle. The mystery fueled decades of terrifying stories. To this day, no single explanation has been proven.

LAKE NESS MONSTER -- THE SHADOW BENEATH THE WATER

(Loch Ness, Scotland)

For centuries, people claimed to see a large, moving shape beneath the surface. Photos and sonar hits kept the legend alive. Science suggests waves, floating logs, fish, and even swimming deer could explain sightings. But the lake is deep, dark, and still full of secrets.

THE SUNKEN CITY OF HERACLEION

(Mediterranean Sea, near Egypt)

Ancient writers spoke of a great port city swallowed by the sea. For years, historians doubted it existed. Then divers found temples, statues, and ships underwater. Legend turned out to be history hiding beneath the waves.

THE BLACK SEA SHADOW FLEET

(Black Sea region)

Sailors claimed ancient ships lay perfectly preserved in deep, dark waters. For centuries this sounded impossible. Modern exploration revealed that deep parts of the Black Sea lack oxygen, preserving wooden ships for thousands of years. The legend was more accurate than anyone expected.

THE BERMUDA TRIANGLE

(Atlantic Ocean between Florida, Bermuda, and Puerto Rico)

Stories claimed ships and planes vanished without explanation. Compasses failed. Crews disappeared. While weather, currents, and navigation errors explain many cases, the sheer number of incidents kept the legend alive. The ocean doesn't need magic to be dangerous.

THE VANISHED WARSHIPS

(Worldwide oceans)

Sailors whispered that some warships simply disappeared without sinking. During wars, chaos and secrecy hid the truth. Decades later, sonar found many of these ships resting quietly on the seafloor. They were never gone -- just waiting.

DID YOU KNOW ?

Did You Know ? Many famous wrecks are left untouched because moving them can actually destroy history. Wood, metal, and even cloth can crumble when brought up after centuries underwater. That's why scientists often study wrecks exactly where they rest instead of lifting them out. Sometimes the safest place for history is the seafloor.

Did You Know ? Freshwater and saltwater preserve objects very differently. Saltwater encourages corrosion and sea life growth, while cold freshwater can slow decay dramatically. That's why airplanes and ships in lakes often look "newer" than those in oceans. The type of water matters as much as depth.

Did You Know ? Some underwater discoveries are kept secret for years before being announced. Governments may delay sharing locations to protect sites from looting or damage. This secrecy can make discoveries seem mysterious or mythical when they finally surface. Silence often protects science.

Did You Know ? Rivers constantly move their bottoms, burying and uncovering wrecks over time. A ship lost centuries ago might be hidden deep under mud, then suddenly reappear after a flood. That's why river discoveries sometimes seem to come out of nowhere. The river decides when to reveal the past.

Did You Know ? Not all sunken objects stay where they land. Strong underwater currents can slowly drag wrecks across the seafloor over decades. Some ships have moved hundreds of feet from their original sinking spot. The ocean quietly rearranges history.

STORY MOMENT

STORY MOMENT

The Screen Lit Up

The research ship rocked gently as waves rolled beneath it, a steady rhythm felt more than heard. Inside the control room, the lights were dim, and every eye was fixed on a single screen. A remotely operated vehicle was descending through dark water, its cameras cutting thin cones of light into the black, revealing only what the ocean allowed for a moment at a time.

At first, there was nothing -- just drifting particles, slow snow made of plankton and silt, and endless empty space. Depth numbers ticked downward. Pressure climbed. The water swallowed sound.

Then something appeared.

A straight line. Too straight to be natural.

The operator eased off the controls. The vehicle slowed, hovering. The shape grew clearer: metal edges, sharp angles softened by time, something long and broken but unmistakable. A ship. Not a legend. Not a guess. A real vessel, resting silently on the seafloor where it had fallen and stayed.

No one spoke.

The cameras glided closer, revealing railings bent by impact, plates peeled open like torn paper, and dark openings where corridors once ran. Sediment lay thick on the deck. Somewhere inside, this ship had carried people with plans, fears, arguments, and destinations they never reached. Watches had stopped here. Journeys had ended here. Now it belonged to the deep.

The ship wasn't famous. No one had written books about it. It had no movie. It wasn't marked on any map. That made the moment even stranger -- more personal, somehow.

This wasn't history found in a museum behind glass.
This was history exactly where it ended.

The crew logged the coordinates, took measurements, and recorded video. There was no cheering. No high-fives. Just quiet focus and a kind of unspoken respect. Some things aren't meant to be touched.

As the vehicle rose back toward the surface, the lights faded and the ship slowly vanished into darkness again, swallowed inch by inch.

It would still be there tomorrow. And next year. And maybe long after everyone on the ship above was gone.

The ocean didn't hide it to be cruel.
It hid it because some stories aren't finished yet.

FUN QUIZ

---

1. **Which famous passenger ship sank in 1912 and still rests deep in the North Atlantic Ocean?**

   A) Lusitania

   B) Titanic

   C) Andrea Doria

   D) Britannic

2. **Why are many shipwrecks left untouched on the seafloor?**

   A) They are too boring to study

   B) They are usually empty

   C) Moving them can destroy fragile history

   D) No one knows where they are

3. **What makes cold freshwater lakes especially good at preserving wrecks?**

   A) Strong currents

   B) Lack of sunlight

   C) Slower corrosion and decay

   D) More sea life

4. **Which type of vehicle is commonly found underwater due to World War II training accidents?**

   A) Tanks

   B) Trains

C) Aircraft

D) Trucks

5. **Why were some military ships and submarines deliberately sunk after wars?**

   A) To hide dangerous technology

   B) To save money

   C) To clear shipping lanes

   D) To create reefs

6. **What often explains sightings of so-called "ghost ships"?**

   A) Supernatural forces

   B) Time travel

   C) Mirages, storms, and drifting vessels

   D) Secret experiments

7. **Why can rivers suddenly reveal wrecks that were hidden for centuries?**

   A) Earthquakes drain the water

   B) Water freezes solid

   C) Floods and shifting mud uncover them

   D) Divers remove the sand

FUN QUIZ ANSWERS

1) B) Titanic

2) C) Moving them can destroy fragile history

3) C) Slower corrosion and decay

4) C) Aircraft

5) A) To hide dangerous technology

6) C) Mirages, storms, and drifting vessels

7) C) Floods and shifting mud uncover them

GAMES

Megalodon Fights Sea Monsters

Google Play

https://play.google.com/store/apps/details?id=com.Dexus.Dinosaur
.MegalodonFights.**SeaMonsters**&pcampaignid=web_share

Prepare to dive into the ultimate underwater battle in this prehistoric sea monster fight simulator! The legendary Megalodon, ancestor of all sharks and king of the deep, has returned to dominate the ancient oceans. This ultimate predator of the sea ventures into uncharted waters, challenging every aquatic creature from the Jurassic, Triassic, and Cretaceous eras.

GAMES

Fishing Carnival - Fish Game

Google Play

https://play.google.com/store/apps/details?id=com.golden.fishing.android.avidly&pcampaignid=web_share

🎉 Welcome to Fishing Carnival! Looking for a brand new casual fishing game? Like to compete with multi-players at a time? Fishing Carnival has got you covered! Join to enjoy the best arcade fishing game online for free!

Act Now - offer may be gone before the tide changes

GET FREE eBOOK(s)
as available

Sign up at our secure website

https://MindBlownBooks.com

Paperbacks, e-Books, Activity Books,

Audio Books + More

THANK YOU FOR EXPLORING WITH US

We're grateful you spent your time with these stories, facts, myths, and legends... and we hope at least one of them made you stop and say, **"No way... that can't be real."**

Curiosity is powerful. When you follow it, the world of any size opens up in ways you never expected. You showed up ready to explore. Awesome!

**Thanks again for reading & wondering.
And for being the kind of person who looks up.**

See you in the next book. **Coming Soon**.

The Official Stamp of Awesome Weirdness

SAVE the
WHALES

savethewhales.org
Founded in 1977